OXFORD GEOGRAPHICAL AND ENVIRONMENTAL STUDIES

General Editors: Gordon Clark, Andrew Goudie, and Ceri Peach

THAILAND AT THE MARGINS

ALSO PUBLISHED BY
OXFORD UNIVERSITY PRESS
IN THE OXFORD GEOGRAPHICAL AND
ENVIRONMENTAL STUDIES SERIES

The Globalized City
Economic Restructuring and Social Polarization in European Cities
Edited by Frank Moulaert, Arantxa Rodriguez, and Erik Swyngedouw

Of States and Cities
The Partitioning of Urban Space
Edited by Peter Marcuse and Ronald van Kempen

Globalization and Integrated Area Development in European Cities
Frank Moulaert

Globalization and Urban Change
Capital, Culture, and Pacific Rim Mega-Projects
Kris Olds

Sustainable Livelihoods in Kalahari Environments
Edited by Deborah Sporton and David S. G. Thomas

Conflict, Consensus, and Rationality in Environmental Planning
An Institutional Discourse Approach
Yvonne Rydin

Social Power and the Urbanization of Water
Flows of Power
Erik Swyngedouw

An Uncooperative Commodity
Privatizing Water in England and Wales
Karen J. Bakker

Manufacturing Culture
The Institutional Geography of Industrial Practice
Meric S. Gertler

Thailand at the Margins

Internationalization of the State and the Transformation of Labour

Jim Glassman

OXFORD
UNIVERSITY PRESS

OXFORD
UNIVERSITY PRESS

Great Clarendon Street, Oxford OX2 6DP

Oxford University Press is a department of the University of Oxford.
It furthers the University's objective of excellence in research, scholarship,
and education by publishing worldwide in

Oxford New York

Auckland Bangkok Buenos Aires Cape Town Chennai
Dar es Salaam Delhi Hong Kong Istanbul Karachi Kolkata
Kuala Lumpur Madrid Melbourne Mexico City Mumbai Nairobi
São Paulo Shanghai Taipei Tokyo Toronto

Published in the United States
by Oxford University Press Inc., New York

First published 2004

British Library Cataloguing in Publication Data
Data available

Library of Congress Cataloging in Publication Data
Data available

ISBN 0–19–926763–4

1 3 5 7 9 10 8 6 4 2

Typeset by Hope Services (Abingdon) Ltd.
Printed in Great Britain
on acid-free paper by
Biddles Ltd.,
King's Lynn, Norfolk

For Oi

EDITORS' PREFACE

Geography and environmental studies are two closely related and burgeoning fields of academic enquiry. Both have grown rapidly over the past few decades. At once catholic in its approach and yet strongly committed to a comprehensive understanding of the world, geography has focused upon the interaction between global and local phenomena. Environmental studies, on the other hand, have shared with the discipline of geography an engagement with different disciplines, addressing wide-ranging and significant environmental issues in the scientific community and the policy community. From the analysis of climate change and physical environmental processes to the cultural dislocations of postmodernism in human geography, these two fields of enquiry have been at the forefront of attempts to comprehend transformations taking place in the world, manifesting themselves as a variety of separate but interrelated spatial scales.

The *Oxford Geographical and Environmental Studies* series aims to reflect this diversity and engagement. Our goal is to publish the best original research in the two related fields, and, in doing so, to demonstrate the significance of geographical and environmental perspectives for understanding the contemporary world. As a consequence, our scope is deliberately international and ranges widely in terms of topics, approaches, and methodologies. Authors are welcome from all corners of the globe. We hope the series will help to redefine the frontiers of knowledge and build bridges within the fields of geography and environmental studies. We hope also that it will cement links with issues and approaches that have originated outside the strict confines of these disciplines. In doing so, our publications contribute to the frontiers of research and knowledge while representing the fruits of particular and diverse scholarly traditions.

Gordon L. Clark
Andrew Goudie
Ceri Peach

PREFACE

Like all books, this one is at least in part an effort to make sense of personal experiences. When I first travelled to Thailand in 1993, the country was in the midst of a long economic boom that led many to speak of Thailand as another Asian 'miracle' economy. At the same time, the horrible Kader factory fire of that year had led many others to highlight the darker side of Thailand's rapid industrial growth. By way of personal relationships, I was immediately immersed experientially in both sides of this contradiction. Many of my connections to Thailand have been through the 1970s student activists whose struggles helped make the country a more democratic place than it once was. These activists themselves have moved on to define different sides of Thailand's divide: some are wealthy, and some occupy leading positions in government; others (though not poor) continue to be leading voices in struggles for the empowerment of the most marginalized groups.

One aspect of my personal experience that I have thus felt the need to explore in this book is the contradictory character of Thailand's remarkable economic growth boom and industrial transformation—at once a successful case of development that has transformed Thailand into a more prosperous society, and yet simultaneously an egregious case of 'maldevelopment' in which the general prosperity and social-political possibilities bequeathed by growth have produced marginalization and even immiseration for far too many.

By the time I had formulated a research agenda designed to address this contradictory process and had returned to the country for field research in 1996, events took a new and dramatic turn. By the end of 1996, and more so by the end of 1997, it was no longer possible to speak uncritically of Thailand as an economic success story because the onset of the economic crisis meant that it was now not just the marginalized who spoke of Thailand's serious development problems but many of the more empowered as well. My interpretive challenge then became not just to analyse the contradictory character of Thailand's developmental 'success' but also its devolution into crisis.

In assessing the 'failure' manifest by the crisis, I have tried to maintain the same sense of contradiction that animates my interpretation of Thailand's 'success'. If rapid growth does not mean prosperity and empowerment for all, crisis does not necessarily mean permanent derailing of the economy. Capitalist development is always marked by contradictions, and crises are 'normal' processes through which these contradictions are both expressed and temporarily resolved. Moreover, neither periods of rapid growth nor periods of crisis should be taken as mere end points but rather as moments in the evolution of social struggles—struggles that are always ongoing. Indeed, if there is anything that I have thus far been able to take from my own experiences of Thailand that can serve as a 'lesson' of the country's contradictory development, it is the primacy of these struggles as the motors of change. I hope that in this book I can breathe some life into this assertion.

Those who have helped me interpret Thailand's social struggles are too numerous to list, and I hope the many that go unacknowledged here will recognize that I am nonetheless grateful for their help. I must add, too, that in spite of the tremendous successes of many struggles for peace, democratization, and social justice in Thailand, the country is still a place where far too many dedicated partisans in struggles for justice meet the worst fates that those with power can arrange. Because of this—and also because my interpretation of power is a critical analysis of institutions and social processes rather than a criticism of individuals—I have chosen to leave most informants and interview subjects unidentified, whatever their particular social position or political orientation.

That having been said, there are a number of people both inside and outside of Thailand who facilitated my research in important ways and to whom I owe a debt of gratitude that requires public acknowledgement. Chayan Vaddhanaphuti and Chaiyan Rajchagool of Chiang Mai University both helped to arrange my stay in Chiang Mai during 1996–7, while sharing with me their considerable knowledge of Thailand. I would also like to acknowledge the hospitality of the Chiang Mai University Social Research Institute during this stay. In Bangkok, Voravidh Charoenloet and Lae Dilokvidhayarat of Chulalongkorn University helped me in special ways to understand some of the struggles of Thai workers and labour organizations. I also owe special thanks to a number of Thai scholars and activists who have shared their work and their insights with me at different points in my research, including Anan Ganjanapan, Chris Baker, Benja Jiraphatpimol, Buddhi Netiprawat, Bundit Thanachaisethavut, Chatchawan Tongdeelert, Chokchai Suttawet, Chusak Wittayapak, Ji Ungphakorn, Medhi Krongkaew, Mingsarn Kaosa-ard, Alain Mounier, Narong Petprasert, the late Nikom Chandravithun, Oraphan Methadilokun, Pasuk Phongpaichit, Prayoon Akarabaworn, Mark Ritchie, Chris Sneddon, Somchai Sirichai, Suchat Trakulhuthip, Sungsidh Piriyarangsan, Sally Theobald, Saowalak Chayataweep, Siwarak Siwarom, Somboon Srikamdokkae, Somsak Tambunlertchai, Surachart Bumrungsuk, Sutham Saengprathum, Thaweerak Thongkontao, Thongchai Winichakul, Uthaiwan Kanjanakamol, Wanida Thantawittayapitak, Wasan Siripoon, and Wilaiwan Saetia. Naturally, the interpretations offered here (and any errors) are mine and not theirs, but the various insights they have shared with me are much appreciated.

I also owe special thanks to people who have read and given me comments on different sections of the book at different stages of its development. These include John Agnew, David Angel, Noam Chomsky, Bud Duvall, Maggi Wai-han Leung, Richa Nagar, Phil Porter, Douglas Webster, and Connie Weil. I am especially grateful to Peter Bell and Katherine Bowie for being frequent interlocutors, as well as inspirational mentors.

I owe a particularly profound debt to Abdi Samatar and Eric Sheppard of the University of Minnesota who not only provided guidance and feedback throughout this project, but also helped make me—and my work, I hope—more geographically sensitive. The MacArthur Program at the University of Minnesota, and its director Allen Isaacman, provided an energizing interdisciplinary environment where much of this geographical thinking could gestate during my time as a MacArthur Scholar. Moreover, both the MacArthur Program and the Geography Department at the University of Minnesota provided financial support during 1994–8, making my research possible. I would also like to thank the Department of Geography at the University of British Columbia and the Izaak Walton Killam Postdoctoral Fellowship Program, which provided support for further research during 1999–2000, along with the Department of Geography and the Maxwell School at Syracuse University, which provided both financial support and a hospitable habitat for completion of the research and writing of the manuscript during 2000–2. Eric Leinberger of the University of British Columbia Geography Department did a wonderful job with the maps. Finally, I would like to thank Anne Ashby of Oxford University Press for her excellent editorial guidance.

An earlier version of Chapter 1 appeared as 'State Power Beyond the "Territorial Trap": The Internationalization of the State', *Political Geography*, 18/6: 669–96, and is reproduced here with the permission of Elsevier. An earlier version of Chapter 6 appeared as 'Economic Crisis in Asia: The Case of Thailand', *Economic Geography*, 77/2: 122–47, and is reprinted here with the permission of Clark University.

The debts that one owes to family and friends when one is undertaking any such extended work as this are both too great to adequately capture in a preface and too important to let go unmentioned. My families in both the United States and Thailand have been patient in enduring—while sometimes fortunately amusing themselves with—my various academic obsessions and projects. Friends, colleagues, and students at the University of Minnesota, Syracuse University, and the University of British Columbia have done much to make the process of producing this book more enjoyable. Most of all, Thitiya Phaobtong has been steadfast in her personal support as well as in her commitments to the social struggles that have animated my research. To her and to all those who still struggle for social justice in Thailand, my love, gratitude, and solidarity.

J. G.

Vancouver
12 August 2003

CONTENTS

LIST OF FIGURES

LIST OF MAPS

LIST OF TABLES

LIST OF ABBREVIATIONS

AAFLI	Asian-American Free Labor Institute
ADB	Asian Development Bank
AFL-CIO	American Federation of Labor-Congress of Industrial Organizations
AoP	Assembly of the Poor
BIR	banking, insurance, and real estate
BMR	Bangkok Metropolitan Region
BoI	Board of Investment
BPP	Border Patrol Police
CIA	Central Intelligence Agency
CLU	Central Labour Union
CMA	Chiang Mai extended metropolitan area
CP	Charoen Phokphand
CPT	Communist Party of Thailand
CSOC	Communist Suppression Operations Command
EPZ	export processing zone
FDI	foreign direct investment
FES	Freiderich Ebert Stiftung
FLUT	Federation of Labour Unions of Thailand
FRUS	*Foreign Relations of the United States*
GDP	gross domestic product
GIZ	general industrial zone
GRP	gross regional product
HDI	Human Development Index
ICA	International Cooperation Administration
ICFTU	International Confederation of Free Trade Unions
IEAT	Industrial Estate Authority of Thailand
IFCT	Industrial Finance Corporation of Thailand
ILO	International Labor Organization
IMF	International Monetary Fund
IPE	international political economy
IR	international relations
ISOC	Internal Security Operations Command
JILAF	Japanese International Labour Foundation
LCCT	Labour Coordination Centre of Thailand
LFS	*Labour Force Survey*
MVA	manufacturing value added
NEDB	National Economic Development Board
NESDB	National Economic and Social Development Board
NGO	non-governmental organization
NIC	newly industrializing country
NPL	non-performing loan
NRIE	Northern Region Industrial Estate
NSC	National Security Council
NSCT	National Student Centre of Thailand
NSO	National Statistical Office
ODA	Overseas Development Assistance
OPS	Office of Public Safety
PARU	Police Aerial Reconnaissance (Resupply) Unit
PFT	Peasant Federation of Thailand

SAP	structural adjustment programme
SDP	Social Democrat Party
SES	*Socio-Economic Survey*
SIFC	Small Industry Finance Corporation
SME	small and medium enterprises
SWNCC	State–War–Navy Coordinating Committee
TAMC	Thai Asset Management Corporation
TDRI	Thailand Development Research Institute
TLU	Thai Labour Union
TNC	transnational corporation
TNTUC	Thai National Trade Union Confederation
TPI	Thai Petrochemical Industry
UNCTAD	United Nations Conference on Trade and Development
UNDP	United Nations Development Program
USAID	US Agency for International Development
USOM	US Operations Mission
VAT	value added tax
WEPT	Work and Environment Related Patients' Network of Thailand
WFTU	World Federation of Trade Unions

Introduction. The Problematic: Territorial State, International Capital, and Uneven Industrial Development in Thailand

The performance of the Thai economy in recent decades has been nothing short of stunning. In 1950, following an entire century of economic stagnation, Thailand was one of the world's poorest countries. Since then, its economy has experienced rapid growth, declining poverty, and macroeconomic stability.

Peter Warr and Bhanupong Nidhiprabha 1996

If Thailand is a success story, it is a highly problematic one. A careful examination may reveal that it is, rather, an exemplary case of 'maldevelopment'.

Peter Bell 1996

What hit Thailand in 1997 was no mere recession but, according to some, a historical cataclysm on a par with the Burmese sacking of the ancient Thai capital of Ayuthaya in 1765–67.

Walden Bello, Shea Cunningham, and Li Kheng Poh 1998

Thailand is one of those export-led success stories that are the best advertisement for the WTO and its free-trade agenda . . . and even post-crisis Thailand is a much better place than it used to be.

Paul Krugman, *Bangkok Post*, 17 Feb. 2000

Whither—and from whence—Thailand?

Thailand presents a vexingly 'hybrid' image of both success and failure. A lay reader of journalistic and academic literature on Thai development could readily be excused for concurring with the opinion of renowned Thai scholar David Wilson, who long ago insisted 'What damn good is this country—you can't compare it to anything!' (Anderson 1978: 193). Indeed, so contradictory are the varying images of the country and the events taking place within it that it often seems Thailand can't even be compared to itself. This sense of identity crisis has only been heightened by the economic crisis that began in 1996. What had been one of Asia's miracle economies led the region into bust, leaving many analysts gasping for air. Thus, as the twenty-first century begins, and we look back on the events of the past century, there is a sense of urgency and contentiousness surrounding a very basic question that one might have already expected to be answered: what exactly is this multifaceted and volatile phenomenon called 'development' in Thailand, and why does it generate such diverse evaluations?

To be sure, some of the contention is due to non-negotiable differences in political perspectives. Yet even granting this, there seems to be less agreement about how to assess development in Thailand than development in many other places. For example, few South Korea scholars, of whatever persuasion, disagree that the country exhibited remarkable and sustained economic growth in recent decades, that this has at least laid the foundations for significant improvements in overall standards of living for most of the population, or that in spite of the crisis the Korean political economy still has substantial potential for further development. Nor, for that matter, do many people disagree that the development process in South Korea was driven forward by an authoritarian state and that issues such as social justice and environmental sustainability must still be addressed.

In contrast, interpreters of Thailand's development experience seem to disagree about such fundamental issues as the importance of state involvement in the process, the degree of well-being which it has bequeathed to the general population, and the future prospects of development. Is Thailand a case of successful development or rather a case of maldevelopment? Is the Thai state a laissez-faire, market-friendly state or an interventionist state on the Northeast Asian model? Is the development process in Thailand sustainable (in economic, socio-cultural, and ecological terms) or unsustainable? Has it improved the prospects for democracy and social justice or has it bequeathed deeper inequalities and lesser likelihood of substantive democracy? And is the economic crisis in which Thailand was mired by the end of the twentieth century the end of the Thai miracle or just a small hiccup on the way to future rounds of rapid growth? All of these questions and many more generate strenuous debate.

I raise these questions not merely rhetorically, for I take a side in the debates. But at the outset it is well to point out that many of the conflicting positions expressed above are themselves reflective of the contradictory totality that is development in Thailand. In other words, there are at least some elements of truth in each of these varied and competing images. Part of the challenge, then, is to analyse the complex, contradictory, and interrelated forces that lead to these images, and to do so in a way that explains why one or another (necessarily partial) view might be recommended over another.

The approach I take to this challenge is to analyse Thai development through three specific, interrelated windows, offering admittedly partial—but nonetheless informative—views: (1) the window of internationalization and uneven development as a socio-spatial process; (2) the window of industrialization and transformation of the labour process; and, most centrally, (3) the window of the state and its role in development. These windows do not define Thai development in its totality, but they provide useful perspectives on many of the crucial, interconnected forces that constitute the development process.

The choice of the state as a primary optic is not due to any endorsement of state-centric explanation. There are a number of other lenses through which one might construct accounts of Thai development, including (for example) the lens of industrial (or firm) structure and the lens of regional and international production networks. My choice of the state as frame of reference is strategic. First, highlighting the role of the state is one way of insisting that economic processes are always simultaneously politically and socially embedded—a point that is widely acknowledged in principle but quickly ignored in practice within much literature on economic development. Second, perhaps no topic has more frequently engaged scholars of recent Asian development than the issue of the state, and thus an analysis of the Thai development process through this lens places the discussion firmly within a broader debate of ongoing interest and significance.

In highlighting the role of the state I do not mean to construe it as an independent variable, nor as a fully autonomous actor in the development process. Rather, the state provides a lens on development precisely in its *relationship* to other facets of this process. Thus, the role of the state as a site of—and organizer of—labour struggles is central to my discussion. Moreover, the state's relationship to the labour process is integrally connected to phenomena such as the development of Thailand's industrial structure and the evolution of socio-spatial inequalities in the process of Thailand's industrialization. In this sense, approaching the issue of Thai development by highlighting labour struggles and the state both prevents crucial class issues from going unanalysed—as too often occurs—and allows issues like those of industrial structure and industrial transformation to be addressed in relation to class processes.

For example, while state influence on the development of industrial structures is often analysed with reference to direct attempts by states to promote particular industries or to abet particular investment decisions, the state's role as a site of labour struggle and an organizer of labour relations has a crucial *indirect* effect on industrial structure that is sometimes missed. Specifically, the strength or weakness of struggles by different groups of workers to appropriate larger shares of the surplus they produce has a major impact on investment decisions, including both locational decisions—as when investors choose to locate either in areas of high wages and market demand or in areas of low wages and production costs—and decisions about technological upgrading and worker training—as when investors replace high-wage workers with machinery or invest in improving employee skills. Thus, the state's involvement in transforming the labour process is interrelated in crucial ways with the development or underdevelopment of higher value-added industries and the concentration or decentralization of industry.

The latter point already indicates the degree to which states and state policies are connected to processes of internationalization and uneven development, and this leads to a final strategic reason for using the state as a lens through which to view Thai development. For while the state has been fairly widely discussed as a historical phenomenon, outside of the discipline of geography there has been less diagnosis of it as a *geographical* phenomenon—a shortcoming I attempt to address here. In this regard, my central goal is to construct a geographical-historical materialist analysis of the Thai state that highlights its role in industrial transformation and uneven development. In particular, I attempt to problematize what could be called the space of the capitalist state in Thailand, and show how this problematization affects the reading that can be given to Thai development.

Internationalization of the state in Thailand

I problematize the Thai state by elaborating and employing a conception of 'the internationalization of the state'. I argue that the internationalization of capital intrinsic to the accumulation process under capitalism involves a transformation of the state that allows state powers to be more effectively exercised by any and all states on behalf of the most internationalized fractions of capital. This internationalization of the state affects not only the ways in which it directly carries out its functions of fostering capital accumulation but also a whole series of state activities that indirectly relate to the production and reproduction of an industrial labour force, including the organization of the labour relations system and the deployment of existing gender relations to enable specific forms of exploitation of women's labour.

Internationalization of capital and the state, as part of the generally uneven pattern of capitalist development, is directly involved in the uneven spatial development of capitalism, both internationally and *within* national territories; and it is also involved in the response of the state apparatus to this uneven development. Internationalization may lead, on the one hand, to concentration of industrial growth in regions strongly connected to the global economy (primarily around primate cities). On the other hand, it may also lead to particular kinds of state efforts to decentralize industrial growth. Moreover, given the gendering of the industrial labour process, there is often an observable gendering of the geography of industrial development, with much of the newly incorporated female working class being more marginal both socially and spatially.

Internationalization of the state and uneven industrial development in Thailand

I employ this conception of internationalization by discussing the development of the Thai state since World War II. To be sure, the Thai state can be seen as having been internationalized over a much longer period of time—indeed, since prior to European intervention in Southeast Asia. Moreover, its internationalization was hastened (and transformed) by British influence in Thailand from the middle of the nineteenth century up until the 1930s. Yet here I deal only with the specific forms of internationalization most relevant to Thailand's recent industrial transformation, a transformation that for the most part dates from after World War II.

The internationalization that occurred during this period was initially enabled by the incursion of the United States into Southeast Asia. Under US hegemony, the Thai peasantry was more fully incorporated into the international market economy, and the project of export manufacturing was initiated, with former peasants supplying much of the industrial labour. The Thai state, acting in coordination with US advisers, developed a wide range of state institutions and projects to further this transformation. In rural areas such as those of Northern Thailand, which I examine in some detail, the state moved aggressively to create local leaders and to incorporate villagers more fully into anti-communist and pro-capitalist market expansion. All of these endeavours facilitated high rates of economic growth throughout the decades after World War II, and set the stage for the even higher growth rates that were to follow. They also encouraged the Bangkok-centric and regionally imbalanced growth that has become a hallmark of development in Thailand.

Since at least 1980, unilateral US influence has receded but a more multipolar internationalization has taken its place, with Japanese investors playing a leading role, often working from behind the initiatives of regional development banks such as the Asian Development Bank (ADB), or with the help of Overseas Development Assistance (ODA) organized by the Japanese state. The Thai economy has been effectively transformed into a full participant in globalized export manufacturing, and rapid industrial transformation has occurred. The state agencies developed under US hegemony have carried out crucial functions in fostering foreign direct investment and manufacturing growth, but have not prevented geographically and socially unbalanced capital accumulation.

One response to these imbalances has been to promote geographical expansion of industrial activity by collaborating with foreign investors in the creation of upcountry industrial estates and export platforms, as in the case of recent manufacturing development in the Chiang Mai region. Measures such as these have helped extend Thailand's rapid economic growth and industrial transformation outward from Bangkok, though without undermining

Bangkok's domination of the national economy. Moreover, the Bangkok-dominated industrial growth process has taken a gendered spatial form, with marginalized women labourers doing much of the production work, both in factories on the outskirts of the Bangkok Metropolitan Region (BMR) and in the even more marginal, and newly emerging, industrial zones of upcountry areas such as Chiang Mai.

Internationalization of the state and social struggle in Thailand

In viewing the state as (unevenly) internationalized, rather than as a merely national entity, I suggest more than just a perspective on the forms and functions of the state and the ways these have facilitated rapid growth and uneven development. Internationalization of capital and the state also frames the process of contestation and state response that has gone on around industrialization in Thailand. I demonstrate this by discussing the development of the Thai Cold War state and both its repression of political opposition and its disciplining of labour. While this repression and discipline took a more overt form in the Cold War period, it has continued to play an important role in the post-Cold War era.

Such state activity has never gone totally unchallenged, however, and to illustrate this I examine recent struggles of Thai workers over health and safety standards in Thai factories. These struggles can be better understood—in both their possibilities and their limitations—by examining their relationship to different segments of the fragmented state bureaucracy, as well as by examining their geographical reach. This reach, or lack of reach as the case may sometimes be, includes the relationships of worker struggles to other social struggles within Thailand, such as those undertaken by peasant and environmental organizations. Specifically, struggles for worker health, led primarily by women, have gained strength from their alliances with various national-level social movements, but they have also been weakened by the limited geographic reach of workers' organizations in Thailand, including the absence of unions from regions such as the North and the weak links of Thai labour with any powerful international labour or political movement—an enduring legacy of the Cold War state's repression of labour and leftist political opposition. Uneven development in Thailand, this suggests, has taken place in a way that disproportionately strengthens internationalized fractions of capital and the most internationally oriented segments of the state, leaving less internationalized groups such as labour at a great disadvantage. In this sense, the inegalitarian nature of Thailand's rapid economic growth is directly connected to the specific forms of its internationalized development.

Such a perspective suggests not only a way of reading recent Thai history and political economy but also an approach to re-framing the aggregate national economic data that are employed in discussions of Thai development. The way in which Thailand's economic growth is related to regional and global developments necessitates viewing Thailand's economic record in relation to—rather than just alongside—the records of other nations. At the same time, a geographical-historical perspective on social struggle and state response points to the importance of foregrounding those sub-national data that reveal uneven development—rather than just adding these on as afterthoughts once the story of rapid economic growth is complete. While such a strategy of re-framing national economic indicators does not satisfy the need for more adequate transnational and sub-national comparative data, it at least allows for interpretations of industrial transformation that can more satisfactorily address both 'internal' disparities in the growth process and the relationship of this process to 'external' forces. In short, it opens the possibility of elaborating integral connections between the internationalization of capital in Thailand and Thailand's contested process of uneven development.

Thailand and East Asian development

Is it the case, however, that David Wilson was right, and that events in Thailand, however interpreted, bear no comparison to those anywhere else? I argue not, and that indeed the internationalization approach to Thai industrialization is useful for analysis of 'late industrialization' more generally—especially throughout Southeast Asia, but perhaps in Northeast Asia as well. In Thailand, the state has played a significant role in the industrialization process, certainly going beyond what is called for in neo-liberal theories of industrial development, but a role quite different from what the 'neo-Weberian' advocates of 'developmental states' have discerned in Northeast Asia.[1] I argue that while the perspectives of neo-Weberian analysts can help us understand some of the specific liabilities of the Thai industrialization process, the dynamics of the Thai case suggest a need to reassess the factors crucial for industrial growth—even, perhaps, in Northeast Asia.

More specifically, the Thai case suggests that if the details of industrial policy stressed by neo-Weberians are nontrivial, they may nonetheless make a less decisive difference in patterns of growth than do a combination of broad geographical-historical circumstances conjuncturally favouring capitalist development. The most important role of the state in contexts that make possible rapid capitalist growth—such as those contexts which emerged in post-World War II East Asia—may well be to position society, polity, and economy in such a way as to both encourage capitalist investment (domestic and/or foreign) and to ensure favourable access to international markets and financial

[1] Throughout the text, I use East Asia to refer to both Northeast Asia (e.g. China, Hong Kong, Japan, South Korea, Taiwan) and Southeast Asia (e.g. Indonesia, Malaysia, The Philippines, Thailand).

assistance. Here, rather than industrial policy, what moves to the foreground is the state's role in successfully disciplining labour and supporting international expansion of capitalist social relations, a project that loomed large in the context of geopolitical challenges posed by militant and popular leftist political movements throughout post-World War II Asia. Moreover, if the specific projects of the state have been altered in the post-Cold War era of 'globalization,' the state's role in disciplining labour and encouraging regional and international economic linkages is nonetheless ongoing.

In sum, then, I propose that viewing the state as itself an (unevenly) internationalized participant in the uneven internationalization of capital places in perspective the apparent paradox that Thailand represents for some a growth 'miracle' while for others it represents an exemplary case of 'maldevelopment'. These two faces of Thai development are not in truth paradoxical but rather interrelated. Thailand has not experienced rapid growth *in spite of* inegalitarianism. Rather, rapid growth has occurred precisely because a highly internationalized state has positioned itself to achieve such growth through the very activities that lead to inegalitarianism and maldevelopment—namely, the repression of labour and of political opposition. In this sense, Thailand's rapid growth has occurred *because of* inegalitarianism and maldevelopment—and the Thai state has had the capacity to foment such growth at least in part because it is not simply 'the Thai state', representing first and foremost the interests of all Thai people, but is rather a very capable representative of internationalized capitalist interests—both 'domestic' and 'foreign'—within Thailand.

The chapters

Chapter 1 establishes in some detail the theoretical framework I deploy throughout subsequent chapters. The internationalization process needs to be seen, I contend, in class terms, but in terms that also unhinge class from its privileged relationship with the national economy. I argue this point by revisiting discussions of the internationalization of capital and of imperialism. While both the literature on 'globalization' and the work of neo-Weberian scholars has tended to downplay imperialism—arguing that it represents a phase of capitalist development now safely behind us—I argue that the persistence of cores and peripheries in the global economy (at multiple scales) preserves utility for the concept. This is only true, however, insofar as the discussion of imperialism is embedded in a theoretically adequate understanding of the internationalization of capital and the capitalist class structure, so that it does not devolve into inappropriate and analytically vapid images of one nation-state dominating another. In arguing for such a conception of the internationalization of the state, I note how this conception differs from the neo-Weberian conception of the 'developmental state', and I briefly preview the ensuing geographical-historical analysis of Thailand by noting that the Thai state does not readily fit into the developmental state paradigm.

In Chapter 2, I begin this geographical-historical analysis by discussing the process of alliance formation between US and Thai elites in the context of the Cold War. The alliance allowed Thai elites and the US national security state to bring potentially or actually unruly peasants and workers more closely under the control of the state and capital, and thus to limit the potential spread of leftist alternatives to capitalist development. In doing so, it laid the foundations for much of the ensuing pattern of uneven capitalist industrialization in Thailand. Chapter 3 examines how the position of labour and the recent history of industrialization can be understood against this broader historical backdrop. In particular, the development of the Thai labour relations system under US tutelage is best understood as an aspect of anti-communist politics. The partial devolution of this system in the era of Japanese quasi-hegemony and global neo-liberalism has left Thailand with a weakly formed set of labour regulatory institutions and a weak and highly fragmented labour movement—phenomena which are enduring legacies of US–Thai anti-communist politics. In Chapter 4, I carry the analysis of these historical legacies further by arguing that the marginalized position of the Northern Thai labour force, and the way it is being incorporated under projects of manufacturing decentralization, need to be understood in relation to prior processes of socio-spatial marginalization. More specifically, I suggest that the marginalization of women factory workers in the North is connected to the gendered spatiality of both industrial development and the struggles of Thai labour.

Chapters 5 and 6 rest in fundamental ways on the geographical-historical analysis carried out in the earlier chapters. In particular, I attempt to show that certain aspects of contemporary Thai development can be usefully interpreted by foregrounding the kinds of internationalization processes that have characterized post-World War II development. In Chapter 5, I argue that specific features of Thailand's recent economic growth—for example, similarities between Thailand's patterns and those evident in other Asian newly industrializing countries, and sub-national variations in the relationship between wage and productivity growth within Thailand—can be understood as manifesting the internationalization of capital *and* the internationalization of the state. In addition, I carry the implications of this one step further by revisiting and critiquing neo-Weberian arguments regarding the state's role in late industrialization, both in Thailand and elsewhere, noting in particular the limits of the developmental state thesis as an explanation of broadly shared Asian regional patterns in economic growth and industrial transformation. In Chapter 6, I suggest how the framework of analysis utilized here can be deployed in analysing the economic crisis that has evolved in Thailand since 1996, as well as in analysing the reaction of the Thai state to this crisis. I conclude that while the crisis in Thailand is itself an expression of the specific forms of internationalization shaping Thai capitalism, it is nonetheless the internationalist forces that seem most poised to take advantage of the crisis and to reassert their dominant position. This outcome is contingent, however, on the outcome of numerous, ongoing social struggles, both within and outside of Thailand, including those that have led in the last few years to the rise of the seemingly more nationalist and populist Thai Rak Thai party. In this sense, I will suggest, Thailand has truly entered the 'post-nationalist' era of capitalist 'globalization'—and also an era in which increasing marginality is increasingly contested.

1

State Power Beyond the 'Territorial Trap': The Internationalization of the State

1.1 Introduction

The 1980s were marked by two seemingly antithetical tendencies in theorizing about states. On the one hand, a strong neo-liberal current connected with the rise of Thatcherism–Reaganism—which was deeply imbued with neo-classical economic assumptions—called into question the power or competence of states, suggesting that the states which governed best were those which governed least. Advocates of this position who attended to Third World development issues were particularly convinced that the rise of East Asian newly industrialized countries (NICs), such as South Korea and Taiwan, constituted evidence that states could best facilitate economic growth and development by maintaining open, export-oriented regimes in which markets were allowed to work unhindered (Balassa 1981; Little 1981; Bhagwati 1988).

On the other hand, by the late 1980s, a school of neo-Weberian scholarship developed in direct response to this neo-liberal approach. Taking issue with the neo-liberals' characterizations of East Asian economic growth, a series of these neo-Weberian scholars showed that state intervention in the economy was far more extensive than the neo-liberals had allowed, and that moreover such interventions seemed to have been successful in fomenting industrial transformation (Evans 1989; 1995; Amsden 1989; 1990; Wade 1990).

The neo-Weberians raised telling arguments and evidence against the neo-liberal position, and it is perhaps a small but significant sign of their success that the World Bank grudgingly acknowledged not only the heavy presence of the state in East Asian industrialization but also some limited efficacy to that presence, especially in the financial sector (World Bank 1993; Amsden 1994; Wade 1996*b*). If this was a victory for the neo-Weberians, however, it may well prove pyrrhic now that the powerful East Asian growth dynamic has been slowed by forces that few states in the region appear willing or able to control. Indeed, and paradoxically perhaps, the more neoclassically inclined now seem to acknowledge the existence of 'strong states' in East Asia and use their existence not to explain economic success but rather to explain the economic crisis that spread through the region during 1997–8.

While the debate being waged in these quarters is important and reflects not only significant theoretical differences between neo-liberals and neo-Weberians but also crucial political differences between major players such as US and Japanese officials, the fact that the debate has so strongly defined the terrain on which discussions of the state occur is unfortunate. Like all debaters, neo-liberals and neo-Weberians share certain assumptions that make their dialogue mutually intelligible—and as is often the case, these shared assumptions can be both deeply problematic and difficult to discern amidst the contention.

In particular, neo-liberal critics of *etatism* and its neo-Weberian defenders share a crucial and problematic assumption regarding the nature of states—namely that states and their powers can for practical purposes be thought of as contained within the bounded territories over which they have formal sovereignty. In addition, many neo-liberals and some neo-Weberians also seem to assume that insofar as the forces of internationalization and/or globalization are increasingly important, it follows that state power is diminished—such power being fundamentally anchored in the nation-state apparatuses being superseded by internationalization and globalization. These foundational beliefs produce varied responses, depending upon precisely how the authors in question view the phenomenon of increased economic integration. For the most adamant neo-liberals (e.g. Ohmae 1995), increased international movement of capital in recent years indicates that states are indeed less and less capable of controlling or efficiently governing markets. Neo-Weberians, also being committed to the notion of state powers as fundamentally national in scale, but wishing to maintain the continued importance of these powers, have responded by insisting that globalization is less real and internationalization somewhat less significant than is typically advertised, with national production and consumption remaining at the core of economic activity. As Robert Wade puts the matter in a paradigmatic statement of the neo-Weberian position (Wade 1996*a*: 61):

> The world economy is more inter-national than global. In the bigger national economies, more than 80 percent of production is for domestic consumption and more than 80 percent of investment by domestic investors . . . These points suggest more scope for government actions to boost the productivity of firms operating within their territory than is commonly thought . . .

Wade has backed this claim by convincingly showing that the globalization rhetoric emanating from business schools and the popular press opportunistically oversells the power of international capital and the weakness of states (cf. Hirst and Thompson 1999). Yet the sort of position that Wade articulates is in some ways defensive. On the one hand, even though it can be shown that the global economic integration currently occurring is not new and has not yet reached astronomical proportions, neo-Weberians nonetheless acknowledge that such integration has increased since the 1960s (Wade 1996*a*: 62–6; Weiss 1997: 7); and if the strongest argument against its importance is only that integration is now no greater than in 1913—when it had reached historically high levels—then the neo-Weberian response has certainly conceded something to the globalization camp.

More tellingly, though, the neo-Weberian response has not yet adequately explained why, if globalization is of limited significance, there should have been the wholesale movement of states—particularly in the Third World—in the direction of policy agendas favoured by neo-liberals (Wade 1996*a*: 69–70; Weiss 1997: 15–16). No doubt states are not as overwhelmed by the power of capital as the globalizers claim, yet they routinely behave as if they are (Biersteker 1995). Indeed, Wade's conclusion that 'In the states of the South we may see a reassertion of the role of the state and even a deliberate step toward disintegration from the world economy for [a] distress-driven reason' seems somewhat unprescient in light of the response of East Asian states to the economic crisis in the region, as will be illustrated later for the case of Thailand (Wade 1996*a*: 88).

In several recent works, on the other hand, Linda Weiss picks up some of the basic themes articulated by Wade and crafts a nuanced response to the issue of state capacity in the era of increased economic integration (Weiss 1997). Weiss defines globalization as 'the creation of genuinely global markets in which locational and institutional—and therefore national—constraints no longer matter,' contrasting this with internationalization, in which 'economic integration is being advanced not only by corporations but also by national governments' (Weiss 1997: 4). She argues that there has only been very weak globalization of economic activity, though strong internationalization, and that state power has not declined but rather its adaptability to new circumstances has become more important (Weiss 1997: 6). Moreover, Weiss argues that 'strong states', rather than being overrun and outmanœuvred by international capital, may actually facilitate internationalization of capital or 'so-called "globalization"' (Weiss 1997: 20–6). In sum, then, Weiss challenges the second of the assumptions shared by many of the neo-liberals and neo-Weberians—namely, that international economic integration leads to the erosion of state power.

While Weiss' views on the role of 'strong states' in promoting global economic integration seem well supported by the evidence from Japan, South Korea, and elsewhere in Northeast Asia, it is still problematic in that it recognizes only 'strong states' as progenitors of internationalization, while relegating states such as the second-tier NICs of Southeast Asia to the backburner, in spite of their very clear role in promoting internationalization and in undergirding high rates of Third World manufacturing growth. This seems in part to be the result of Weiss wishing to differentiate the Southeast Asian NICs from the original four East Asian 'tigers' in order to marginalize the implications of the Southeast Asian economic crisis for her discussion (Weiss 1997: 4–5).

But even more to the point, perhaps, is that Weiss' linking of only the 'strong state' with internationalization conforms to the neo-Weberian view that states can or should act primarily in response to domestic classes and social groups. The Northeast Asian states can be seen as promoting internationalization primarily as part of a strategy to strengthen the position of their own domestic constituents—particularly their capitalists, to a lesser extent other groups in society—but this is a characterization that would not so neatly fit the patterns of internationalization in Southeast Asia (Schmidt 1996; Jomo 1997). Thus, even in what is in fact a very powerful rebuttal of neo-liberal claims about the powerlessness of states, the assumption that state power is most fundamentally linked to processes and social groups within the national territory retains its tenacity.

In this chapter, I lay out theoretical groundwork for a different approach to this issue. I suggest that it is true *both* that internationalization of capital is increasingly important *and*, as Weiss and others argue, that states retain a crucial role in the international political economy. But unlike Weiss, I argue that what makes both phenomena simultaneously possible is not the strength or adaptability of national states so much as 'the internationalization of the state' (Cox 1987; Panitch 1994).[1] I define internationalization of the state as *a process in which the state apparatus becomes increasingly oriented towards facilitating capital accumulation for the most internationalized investors, regardless of their nationality*. This process is a dimension of broader internationalization tendencies which, while not erasing national boundaries in the fashion predicted by the globalization literature, has nonetheless created a set of elite-based transnational alliances that strengthen the possibilities for internationalized capital accumulation based less on national priorities than on shared transnational class interests (Sklair 2001). Internationalization of the state thus forces social theorists to problematize the conceptions of the state that have been at the core of the neo-liberal/neo-Weberian debate. In particular, it demands that we think about states in a fashion that is at once more class-based—including breaking down the sort of distinction between politics and economics permeating both neo-liberal and neo-Weberian discourse—and more geographically nuanced—breaking down conventional distinctions between the national and the international.

1.2 Beyond the nation-state container

The notion that nation-states should not be identified fundamentally or exclusively with national territories or nationalist economic policies has been articulated fairly frequently in recent work by political and economic geographers

[1] Weiss refers to the 'internationalization of state capacity' (1997: 4), but does not elaborate a conception of the internationalization of the state in the sense intended here.

(Notermans 1997; Herod, Ó Tuathail, and Roberts 1998; Yeung 1998). In his discussion of transnational corporations and nation-states, for example, Peter Dicken notes the internationalizing activities of states (Dicken 1997). He points out that states, much like firms, compete with one another to gain material advantage, and that in doing so they 'increasingly engage in strategic alliances with their competitors'—as for example in the promotion of regional free trade areas, customs unions, common markets, and economic unions (Dicken 1997: 84–5). Thus, states operate transnationally, in certain respects, much as do corporations. But while this point helps loosen the conceptual connection typically drawn between the state and its formal territorial space, neither Dicken nor other authors who have written in this vein fully investigate the possibilities suggested by such a loosening.

Here, it is helpful to revisit a theoretical argument advanced by John Agnew and Stuart Corbridge in their critique of international relations (IR) theories and International Political Economy (IPE) (Agnew and Corbridge 1995). Agnew and Corbridge present three interrelated geographical assumptions that have informed most varieties of IR theory.[2] These assumptions are: (1) national spaces are fixed units of secure sovereign space; (2) a clean distinction can be made between the domestic and the international; and (3) the territorial state exists prior to and is a container of society (Agnew and Corbridge 1995: 83–4).[3] In arguing for a reconstruction of geopolitical thought informed by IPE, Agnew and Corbridge note that such assumptions are made problematic by a number of contemporary economic realities—for example, by the fact that as much as 40 per cent of what is counted as world trade may actually be movement of goods *within* the same firm, making any attempt to draw clean lines between what is internal and external to a national economy difficult at best (Agnew and Corbridge 1995: 91).

For Agnew and Corbridge, analyses that hew narrowly to the accepted practices of national accounting, without recognizing how internationalization of capital is breaking down some of the boundaries between the national territory and its exterior, are examples of thinking that falls into the 'territorial trap'. Yet it is important to emphasize that the internationalization of capital, which is making the idea of a national economy increasingly problematic, is less a persistently observable empirical phenomenon than a deeply ingrained structural *tendency* of capitalist development. Internationalization is an uneven process that can be counteracted by various forces at particular points in time—e.g. war, economic crisis, etc.—and thus may not be immediately manifest as an empirical reality; but it is nonetheless a perpetual underlying urge of the capitalist accumulation process.[4]

[2] The original version of the argument is in Agnew (1994).

[3] A related but distinct critique of nation-states as containers is developed by Taylor (1994; 1995).

[4] This is a point long recognized in the Marxist political economy literature. See e.g. Murray (1975); Palloix (1975); Harvey (1982; 1985); Jenkins (1987); Storper and Walker (1989).

I mention here several major points regarding this internationalization of capital—points that underpin the conception of internationalization of the state discussed below.

(1) The internationalization of capital is the internationalization of the classes in capitalist society. This means, in particular, that capitalists and urban industrial workers become more globally ubiquitous (Harvey 1982). Moreover, production processes and class relations involving these groups are in the process of actively 'stretching' across space (Storper and Walker 1989; Massey 1995), not infrequently across international boundaries. This can take the form of increased exploitation of a 'national' labour force by 'international' capital, but it can also take the form of increased communication and coordination of activities between workers of different nationalities (Herod 2001) and between capitalists of different nationalities (Shoup and Minter 1977; Sklair 2001).

The latter of these phenomena is of particular significance to the post-World War II process of internationalization, since the US government has acted—through foreign assistance, international training programmes, and other measures—to create a like-minded international business and governmental elite (Cox 1981; 1987; Biersteker 1995). Thus, crucial positions within the upper echelons of both the capitalist class and the capitalist state are increasingly peopled by groups with similar ideologies and class interests (especially in increased international trade and investment). Such elites often share not only an abstract worldview but a common language (English), common technical skills and training, and common consumption norms that set them apart from most inhabitants of the Third World (Jones 1998; Faust and Nagar 2001). Thus, the fact that the majority of smaller capitalists and labourers remain more place-bound and provincial in outlook—corresponding no doubt to the statistical significance of domestic economic activity cited by neo-Weberians—cannot be taken as evidence that capitalism is tightly wed to national strategies of accumulation. Indeed, it remains clear that capitalism is perhaps distinct from all other known forms of socio-economic organization in the intensity of its internationalizing tendencies—and these tendencies are generally most developed within the most powerful fractions of the capitalist class, who have gained the most from past rounds of internationalization.

(2) Internationalization of capital, as a long-term tendency, has been directly connected with the phenomenon of imperialism. This was more transparent in earlier periods of internationalization, when colonialism was the preeminent form of imperial control. The colonial model, however, is misleading as a general model of imperialism in that it foregrounds the national dimensions of domination, when these have been increasingly subordinated to the class-relevant dimensions of imperial projects. Contemporary theories of imperialism (e.g. Baran, Sweezey, Frank, Amin, Wallerstein) have framed the discussion in terms of core–periphery relations, and while this geographic imagery can be instructive it can also mask such class-relevant processes (Brenner 1977; Harvey 1982).

Imperialism is most usefully seen as an expression of capitalism's uneven development (Smith 1991), in which more powerful accumulation processes in capitalist core areas allows the formation of alliances linking the interests of both core and peripheral area elites with those of core area workers (Galtung 1971). In such alliances, a stratum of peripheral elites and a larger and more complex group within core areas benefit from the 'super-exploitation' of workers in peripheral areas (Amin 1974; Blaut 1987; Sheppard and Barnes 1990). Imperial alliances rely heavily on the collaboration of peripheral elites who both benefit from and endorse a capital-accumulation process linking core and periphery (Pastor 1987; Jones 1998)—even if in the form of 'dependent development' (Cardoso and Faletto 1979; Evans 1979). These 'liaison groups' (Petras 1978), or 'transnational kernel' (Sunkel 1973; Evans 1979), are frequently the most dynamic 'local' forces promoting industrial growth within the periphery. Thus, while their performance in helping give rise to Third World manufacturing development can be seen as undermining assertions of perpetual 'underdevelopment' in the global periphery (Frank 1966), their relationship to internationalization processes suggests that the development they help foment cannot be properly regarded as a narrowly national phenomenon. At the same time, the participation of core country elites in these alliances does not typically make core countries or states as dependent on transnationalized accumulation as peripheral states. Thus, imperialism expresses the unevenness of capitalist internationalization.

(3) Internationalization of capital is not merely an economic phenomenon but implies a much broader set of social transformations. Imperialism, for example, is not simply a matter of economic domination but has crucial and interrelated political, cultural, and ideological dimensions that affect both the general structures of the dominated societies and their patterns of capital accumulation (Petras 1978; Slater 1989). In addition, reproduction of capitalist social relations does not occur mechanistically, and so the process of spreading capitalist relations throughout the periphery requires a complex struggle to find appropriate 'modes of regulation' for an increasingly internationalized process of accumulation (Jessop 1990*a*; Overbeek and van der Pijl 1993).

Moreover, social reproduction in the broadest sense demands the explicit or implicit negotiation of 'gender contracts' (Duncan 1994) that allocate various responsibilities between men and women, and these contracts both affect and are transformed by capitalist internationalization (Deere 1976; Benería 1979; Nash and Fernández-Kelly 1983; Standing 1989). Such internationalization can be seen as part of a broader process in which women are drawn increasingly into the wage labour force and begin to exchange more 'private' and household-based forms of subordination to men for more 'public' or workplace-based forms (Brown 1981; Walby 1990; Tiano 1994). At the same time, this movement from private to public patriarchy frequently involves important geographical dimensions, with the development of feminized workplaces typically involving various forms of socio-spatial marginalization (Nelson 1986; Cravey 1998).

In sum, then, internationalization of capital is a complex, multifaceted tendency that is not always manifest and is never uniform but which is nonetheless regularly expressed in actual development processes, especially within the global periphery. Most significantly for the discussion that follows, it has helped generate a highly transnationalized elite, dominated (though not peopled exclusively) by men. It is the presence of these elites within a broader alliance of class forces that helps give a shape to internationalization more palpable than what is betrayed by aggregate economic data.

1.3 The capitalist state

There is by now an enormous literature on the capitalist state, written from a variety of perspectives, and I will not attempt to review this literature here.[5] I start from the assumption that class relations and the capital accumulation process have crucial constraining and enabling effects on the forms and functions of capitalist states. Yet these class forces do not mechanistically produce the sort of capitalist state needed for the reproduction of capitalist social relations and so the fit between the forces at work within the accumulation process and the specific activities of the state is always somewhat loose and underdetermined. For this reason, we can speak of the state 'as the site of class (-relevant) struggles and contradictions,' and also as 'the site of struggles and rivalries among [the state's] different branches' (Jessop 1990*b*: 261).

The state, as this suggests, is not a unitary or homogeneous entity but rather is an ensemble of different institutions among which there is enough contention to insure that there is no simple, mechanical production of the kinds of policies favoured by any one element of the capitalist class. As Bob Jessop puts it, 'The state is a specific institutional ensemble with multiple boundaries, no institutional fixity and no pre-given formal or substantive unity' (Jessop 1990*b*: 267). But this lack of institutional fixity is not merely the result of some loosely conceived relative autonomy of the state or of the actions of individual state managers vying for power through the use of state mechanisms.[6] Rather, it is inherent in the nature of state power, since the state has no permanent or unconditional power of its own but only a set of institutional capacities (Jessop 1990*b*: 269–70):

[5] A fairly comprehensive review of different positions can be found in Jessop (1990*b*). For a discussion of some specificities of Third World capitalist states, see Glassman and Samatar (1997).

[6] State managers can be argued to have an interest as managers in specific outcomes such as limiting inflation (Notermans 1997; Brenner 1999). But the ability of state managers to act on such interests is heavily conditioned by their ability to mobilize large blocs of capital behind their agendas, since capitalist states do not control most of the investment decisions necessary for economic stability and are thus readily subject to destabilization by capital strike (Block 1977).

> . . . the power of the state is the power of the forces acting in and through the state. The forces include state managers as well as class forces, gender groups as well as regional interests, and so forth. State power also depends on the forms and nature of resistance to state interventions—both directly and at a distance from the state.

This sort of 'strategic-relational' view of the state—in which state power is seen as being 'structurally coupled' with the forms of power and struggle emanating within the broader society, rather than as representing any permanent congealed power in its own right—is not incompatible with the notion that state activity is crucial to the reproduction of capitalist society. Indeed, as Jessop's work argues in great detail, it is necessary for capitalist classes to struggle through the institutions of the state to create the conditions for reproduction of capitalism since these conditions cannot be guaranteed solely within the market. State functions are not optional but are in fact necessary for capitalism. But there is no guarantee that the particular functions capitalism as a whole needs for its reproduction will come into being through the state because the state is not a unitary whole in which plan-rational bureaucratic activity of the sort favoured by neo-Weberians readily takes place. Rather, the state internalizes struggles within the broader society and thus also internalizes society's fractiousness. Consequently, the actual activities of the state do not depend narrowly on the forms and functions of the state at any given time but rather on how these are produced and reproduced through social struggle. This shifts the analysis of the state not only to an examination of the dialectic between social forces and state forms and functions but to the strategies adopted by different social actors in struggles over the capital accumulation process (cf. O'Connor 1973).

1.4 Class fractions and the internationalization of the state

In a formative discussion regarding the internationalization of capital, Robin Murray makes several important, general points (Murray 1975). Murray starts from the observation of a 'territorial non-coincidence' between the economic reach of international firms within the global economy and the political reach of nation-states that have economic responsibilities vis-à-vis these internationalized fractions of capital (Murray 1975: 107). The issue that needs resolution here, for Murray, is how the economic functions capital needs the state to carry out can in fact be carried out, given this non-coincidence. For Murray, the extension of capital across international boundaries makes possible more complex relationships between capital and the state than what is conceived to be the case when capital is assumed to have a nationality (Murray 1975: 119):

> When any capital extends beyond its national boundaries, the historical link that binds it to its particular domestic state no longer *necessarily* holds. A capital which has

extended itself in this way will require the performance of the primary public functions for its operations. But the body which performs them need not be the same as the body that performs them within the area of the capital's early development.

Murray identifies five possibilities here. First, the domestic state can perform the functions for overseas capital through geographic expansion, as with colonialism. Second, the domestic state can arrange for foreign structures to carry out the functions through neo-colonial policies. Third, internationalized capital may perform the functions itself, either singly or in conjunction with other capitals. Fourth, foreign states may already be performing the necessary functions or may be willing to do so of their own accord. And finally, existing state bodies may carry out the necessary functions in cooperation with one another (Murray 1975: 120–2).

A central feature of the internationalization of the state is thus the negotiation and coercion used to establish the precise roles of various states in carrying out these economic functions.[7] The internationalization of capital, in short, does not imply that capital transcends the state. In Nicos Poulantzas' words, 'internationalization of capital neither suppresses nor by-passes the nation states' (Poulantzas 1975: 73). Instead, internationalization implies that the narrow association of the most internationalized fractions of capital with particular nation-state apparatuses does not hold and thus the process of determining how necessary state functions will be carried out becomes more complicated. As Poulantzas puts it (in referring to the relationship between US and European capital specifically):

> . . . capital which overflows its national limits certainly has recourse to national states—not only to its own state of origin, but to other states as well. This gives rise to a complex distribution of the role of the state in the international reproduction of capital under the domination of American capital, which can lead to the exercise of the state functions becoming decentered and shifting among their supports, which essentially remain the national states. According to the conjuncture, any one or other of the metropolitan national states may assume responsibility for this or that international intervention in the reproduction process, and for the maintenance of the system as a whole. (Poulantzas 1975: 82–3)

Recognizing 'territorial non-coincidence' between the reach of different states and different fractions of capital thus gives us some purchase on how and why states become internationalized. But since capital is not homogeneous and has different fractions with differing geographic ranges, we cannot leave the matter at this level of generality. Richard Bryan's important discussion of the internationalization of capital and the role of the state (1987) identifies four different types of capital, each related in specific ways to the circuits of capital

[7] A possible criticism of Murray's approach can be forestalled here by citing Panitch (1994: 65): 'To speak in terms of functions is not necessarily improperly "functionalist" insofar as the range of structures that might undertake their performance, and the conditions which might mean their non-performance, are explicitly problematised. Murray explicitly did this . . .'

discussed by Marx—the circuits of money, commodities, and productive capital (cf. Webber and Rigby 1996: 454–7). First, there is capital which, regardless of its formal 'nationality', not only produces commodities within a given national space (this being true of all forms of capital) but which must realize value by selling products and reinvesting in production within that same national space. This fraction of capital, which is historically associated with import-competing industries, Bryan refers to as the 'national' fraction. Juxtaposed to this is a fraction of capital that although producing within a nation-state can sell on global markets and reinvest in production beyond the borders of the nation-state in which the original production process took place. This fraction, for which the transnational corporation (TNC) would be the paradigm, is referred to as 'global' capital. But there are two other fractions that can be delineated. One is capital that Bryan calls 'investment-constrained'. This is capital that can sell on international markets but cannot consider international production because it is not large-scale enough. A final fraction of capital is capital that is 'market-constrained'. This is capital that can invest internationally but can only sell the products thus produced within the national markets where those products are produced. This may again be a circuit for which TNCs are paradigmatic, but in this case the TNCs do not invest in order to export but rather to compete on the domestic market (Bryan 1987: 264–6).

The point in making these sorts of distinctions is that it foregrounds how different fractions of capital systematically benefit from different kinds of state policies. Moreover, these interests themselves are in part shaped in advance by the existence of state policies. Thus, a period in which import-substitution policies are dominant will tend to be identified with the dominance of national and market-constrained capital. By contrast, a period such as the present, where states are undoing many barriers to specific forms of capital movement is associated with the rising importance of global capital, possibly in collaboration with investment-constrained capital (Bryan 1987: 271).

The utility of identifying fractions of capital in this way is that despite the oversimplicity—for example, many TNCs produce both for local markets *and* for export markets—the fractions thus illustrated highlight the structural basis for collaboration between fractions of capital across national territories. For example, both investment-constrained domestic capital and global capital can benefit from state policies promoting exports. On the other hand, TNCs that invest in order to produce for local markets may be comfortable with import-substitution policies that are also favoured by market-constrained domestic producers. In short, the positioning of fractions of capital leads to possibilities for alliances transcending national boundaries and which are grounded in objectively similar interests in particular kinds of state policies. State policies may therefore simultaneously serve certain 'domestic' and certain 'foreign' fractions of capital. States can thus—without being merely subordinate to foreign interests—function to facilitate capital accumulation by foreign nationals within their own national territory.

Another structural support for international alliance-building within the state can be explained by further elaboration on the relationship between the money circuit, the commodity circuit, and the productive circuit. Circulation of capital necessarily involves all of these circuits, but the spatial relationships between fractions of capital identified with each shape the character of the accumulation process in crucial ways. Rob Steven (1994; 1996) argues that it is the interrelationships between these three fractions—and in particular the strength and interpenetration of each—that determines the strength of a given national accumulation project. For Steven, for example, the power of contemporary Japanese imperialism is a function of the *keiretsu* networks and the way they interweave banking, trading, and productive capital in order to extend their activities across different geographical zones.

This perspective provides insight into why Third World states should frequently find alliances with TNCs and imperial states to be both necessary and useful. In general, the uneven development of capitalism has meant that for many developing countries the fractions of capital are unevenly developed. For example, many Third World countries are weak in the development of productive industrial capital, though the most underdeveloped economies may also have weakly developed trading and banking sectors. Since successful capital accumulation requires that all three fractions of capital be well-developed, Third World states and their 'domestic' capitalists frequently find it necessary to build alliances with the more advanced sectors represented by TNCs and imperial states (Cardoso and Faletto 1979; Evans 1979). Given this structural affinity of certain fractions of capital within the Third World for TNC capital, it is not surprising that Third World states should often court TNC investment.

1.5 Fractions of capital and state institutions

This conception of the relationship of fractions of capital to the orientation of the state can now be expanded through examination of concrete ways in which capital shapes specific institutions of the state. Here, much work has already been done by authors looking at the development of the imperial state apparatus—and particularly at the United States. James Petras and Morris Morley (1981: 15–16) argue that certain specific institutions associated with specific fractions of internationalizing capital have played the most significant role in the formulation of imperial state policies. The imperial state apparatus is fractured in several ways: first, between agencies representing the interests of more specific blocs of capital (e.g. the Department of Agriculture) and those representing the more general interests of capital as a whole (e.g. the State Department); second, between agencies closely tied to specific blocs of capital that may be competing with one another (e.g. the Commerce and Treasury Departments); and third, between agencies that have differing types of functions (e.g. the Department of Defense and the US Information Service). There are not always strict dividing lines between institutions based on these sorts of categories—for example, military institutions may carry out economic, military, and ideological projects simultaneously. Nor do the specific divisions of constituencies or responsibilities prevent collaboration between institutions at crucial points in foreign policy processes. But the sorts of distinctions drawn here provide a useful first cut for analysing how specific blocs of capital might end up competing, negotiating, and collaborating with one another, within the institutions of the imperial state, over the orientation of foreign policies.

What can be said here of the internal complexities of peripheral states engaged in relations with states like those described by Petras and Morley? Robert Cox argues that globally the most internationalized segments of capital have been most powerfully represented within specific sections of the nation-state apparatus such as ministries of finance and prime minister's offices, which as Cox notes 'are key points of adjustment of domestic to international economic policy'. By contrast, Cox argues, 'Ministries of industries, labour ministries, planning offices, which had been built up in the context of national corporatism, tended to be subordinated to the central organs of internationalised economic policy' (Cox 1981: 146).

Using Bryan's terminology, here, we might suggest that 'global' fractions of capital, which sell on global markets and reinvest internationally, will be likely to have strong connections within particular segments of the state (such as commerce and finance ministries) that are crucial for the internationalization of goods and reinvestible surplus. The global fraction here would include both 'foreign' TNC capital and 'local' capital that sells and invests abroad. This is not to say, either, that other fractions of capital might not fight for control within such agencies. But since these less-internationalized fractions have perhaps more stake in policies other than deregulation of money capital—for example, in industrial policies that would benefit those producing goods for local markets—they may expend less effort on control of the most internationalized ministries.

Placing this description of the orientation of specific ministries in the context of interstate relations, Cox characterizes the interactions between leading states within the global political economy as centring around a hierarchically structured process of consensus formation between state actors that involves the adjustment of internal state structures 'so that each can best transform the global consensus into national policy and practice, taking account of the specific kinds of obstacles likely to arise in countries occupying the different hierarchically arranged positions in the world economy' (Cox 1987: 254). The fact that the system is hierarchical is crucial, because while consensus can be relied upon to generate the internal adjustment process at the top of the system, coercion and the exercise of cultural imperialism are more prevalent in the relations between state apparatuses at opposing poles of the global hierarchy. As Cox puts it in discussing Third World states (Cox 1987: 260–1):

The top-level countries in effect jointly fix the parameters of the developmental options of late-industrializing countries. Third World elites do not participate with the same effective status as top-level elites in the formation of the consensus. The consensus does, however, gain ideological recruits and places ideologically conditioned agents in key positions within Third World countries. The networks through which international finance flows to these countries are staffed within these countries (e.g. in top positions in central banks and finance ministries) by people who have been socialized to the norms of the consensus and of its professional cadres. These people are often graduates of major advanced-capitalist-country universities and have often passed through the International Monetary Fund (IMF) institute and similar bodies that bring Third World technical financial personnel into personal contact with the milieu of international finance.

Despite the presence within Third World states of such a Western-trained technocratic elite, however, '[t]he internationalizing of the Third World state is more openly induced by external pressures than the internationalizing of the advanced capitalist state is and thus provokes more awareness and resentment' (Cox 1987: 265). Consequently, various types of force have been fairly routine handmaidens of the internationalization process within Third World states.

1.6 Imperialism and the internationalization of the peripheral state

The account of the internationalization of the state developed so far suggests the need to foreground imperialism and its affects on the forms of peripheral states. Imperialism, to reiterate, is not a matter of one nation-state dominating another, but neither is it a process of particular blocs of capital dominating a nation-state. Rather, it is a complex process by which particular fractions of capital, acting through particular branches of the imperial and peripheral state, act to facilitate the forms of internationalization of capital most relevant to their interests. This, again, involves strategic manœuvring and class (-relevant) struggle, enacted through coalition building both outside and within the state.

Historically, one of the major sites of strategic manœuvring has been military aid and the collaboration of military forces that allows the military agencies of imperial states access to the territories of peripheral states. Thus, among the forms of interstate collaboration that are part of imperialism and the internationalization of the state is the direct representation of core area military forces within the peripheral state—these core area military forces not simply representing narrow, militarily defined interests but rather broader societal and class interests that have crystallized within the projects of the core country military. In short, the peripheral country's military is itself one of the state institutions potentially internationalized in specific ways as part of the internationalization of capital within an imperial state system. This means that in spite of the pretence that the peripheral state's military is the national institution par excellence it is in fact a significant site of internationalization.

More recently, however, direct use of military force as a tool of imperialism has become somewhat less imperative than in earlier eras, even though it is still a crucial underpinning of imperial power. This shift manifests the increasing hegemony of global capital and the support for it within collaborator governments of the periphery, in part a reflection of the power such capital is able to exercise through various transnational institutions such as the IMF and World Bank (Cox 1981: 143). An important aspect of global capital's increasing hegemony is the pervasive conformity of most state and business elites throughout the world to an ideology of 'transnational liberalism', which in and of itself makes the direct use of military force on the periphery somewhat less necessary (Agnew and Corbridge 1995; Agnew 1998). In this sense, transnationalized ideologies of competitiveness and minimal state regulation have become an important material force in their own right (Biersteker 1995).[8]

In this context, we should note the ways in which imperialism as a political and cultural phenomenon complicates the picture of the nation-state as a site for the reproduction of national culture. Benedict Anderson's account of the development of nationalism argues that the expansion of the colonial state and the production of a group of colonial subjects educated in colonial schools but excluded from corporate boardrooms 'meant that to an unprecedented extent the key early spokesmen for colonial nationalism were lonely, bilingual intelligentsias unattached to sturdy local bourgeoisies' (Anderson 1991: 140). The results of the colonial educational system, in sum, were paradoxical because they produced subjects identifying with certain core values of the colonizing countries, yet unable to advance on their own within the colonial system.

Part of what distinguishes imperialism in the *post*-colonial period from earlier variants is the shifting character of this peripheral cultural-ideological complex as certain of the older barriers to advancement have fallen. Post-colonial imperialism has successfully created a 'counter-nationalist' intelligentsia, which though still needing to some degree to have national credentials in order to serve within the nation-state has nonetheless taken on board most of the perspectives of its international mentors. Particularly significant here are various economists and others trained in economics, who have frequently had significant influence precisely within some of those peripheral state agencies identified by Cox as crucial to the internationalization of the state (Bell 1991; Biersteker 1995). This counter-nationalist intelligentsia is in effect a 'transnational kernel' within the technocracy and a crucial player in the process of alliance formation that promotes capitalist internationalization. This

[8] In calling ideology a material force, I am following the sort of analysis laid out by Louis Althusser (1972).

suggests, then, that the collaboration and alliance-building crucial to the imperial system transcends narrowly economic class coalitions not only through the internationalization of military force but also through the internationalization of certain political and cultural values and ideals—these again being propagated through complex processes of struggle within the arena of the state (cf. Escobar 1995).

Capitalist internationalization thus involves the state as an internationalizing agent not only vis-à-vis economic processes but vis-à-vis the broad panoply of human activities that constitute social life, since most if not all of these are at some points related in crucial ways to the production and reproduction of capitalism. It is in this sense that Peter Bell, writing on Thailand, refers to capitalism as turning society into a 'social factory', where all aspects of life become more or less conditioned by the requirements of generalized commodity production and attendant ideologies of competitiveness (Bell 1978; cf. Tronti 1971). Within the core of the global economy, this process is highly advanced, so that rather than being merely neutral or apolitical, virtually all aspects of education and popular culture are to some extent politicized in a fashion that is more or less conducive to the social reproduction of class privilege. This is quite important when considering the impact of Western training and technical aid missions upon the development of the bureaucracy in peripheral countries. Even within the realms of the state typically seen as less beholden to the immediate logic of international capital accumulation, ideologies of competitiveness, respect for capitalist property, and so forth have often taken deep root, making advocates of particular forms of education, medicine, labour legislation, or other social services conscious or unconscious supporters of the project of capitalist internationalization.[9]

1.7 Struggle over the internationalization of the state

It is important to emphasize that the internationalization of the state is, like the internationalization of capital, a process in motion. As Cox puts it, 'the tendency toward the internationalizing of the state is never complete, and the further it advances, the more it provokes countertendencies sustained by domestic social groups that have been disadvantaged or excluded in the new domestic realignments' (Cox 1987: 253). The fact that the state is not given as internationalized but is *being* internationalized—and unevenly at that—suggests the need to look for the effects of particular struggles over the forms and functions of the state. What can be identified, if the state is examined in this way, are struggles that correspond to the evolution of the social formation and

[9] Though I follow Bell in adopting Mario Tronti's notion of 'social capital' (Tronti 1971), the perspective being argued for here is more Gramscian than the 'autonomist' Marxist approach favoured by these authors.

the rise or decline of particular fractions of capital—but the specific dynamics of which depend upon concrete political mobilizations by classes and class fractions.

For example, as Philip McMichael and David Myhre note, structural adjustment loans (SALs) have been used by the IMF and World Bank to force a more 'outward' reorientation of economic policies in the Third World—a reorientation 'which facilitates the redistribution of power within the state from programme-oriented ministries (e.g. social services, agriculture, education, etc.) to the central bank and to trade and finance ministries . . . Thus, policymaking is displaced upward from South to North' (McMichael and Myhre 1990: 66). Such displacement, however, has not occurred without contestation, including frequent resistance to IMF measures by popular organizations. This suggests that we should not take the positioning or orientation of particular ministries as given but should rather see internationalization as an ongoing struggle over forms and functions of the state and, within this, the orientations and relative empowerment of various ministries. In addition, we cannot assume the main lines of battle to always be drawn simply *between* ministries. The training of a wide variety of professionals abroad, in conjunction with the development of actively internationalist or expansionist classes in the periphery, means that there are strong tendencies towards internationalization of the state at work within almost any given segment of it, and representatives of these tendencies often fight older or more established interests for control within the ministries. Leo Panitch makes this point when he suggests that ministries and approaches associated with a 'competitiveness ideology' are gaining status and that social welfare agencies 'are perhaps not so much being subordinated as themselves being restructured' (Panitch 1994: 72)—often from within as much as from without.

Given the pervasiveness of recent World Bank and IMF-led attempts to restructure Third World states along neo-liberal lines, and given what this reflects about the tremendous power of capital based in core countries, it is important to conclude by reiterating that the core–periphery features of restructuring do not prevent it from serving the interests of peripheral elites. Thus, the policy orientation that has so repeatedly emerged within Third World states in the post-Fordist era can be seen as reflecting the power of an alliance involving both core and peripheral elites who are closely connected to internationalization processes and dominant within the ministries most aggressively promoting such processes. If this represents a change from the forms of transnational alliance dominating the Fordist period of imperialism, it does not represent the dawning of a post-imperial era (cf. Becker and Sklar 1987). Rather, it simply represents a transformation in the specificities of the alliances and social struggles that constitute contemporary core–periphery relations.

1.8 Developmental state or internationalization of the state? Previewing the Thai case

On the basis of the sort of theoretical account outlined here, I suggest that one can move a step beyond the argument that states *promote* internationalization and argue that states themselves can (and have) *become* increasingly internationalized—that is, oriented at the outset, through their increasingly internationalized social base, towards promoting internationalization of capital. Indeed, it is perhaps the case that so far the most important dimension of the multifaceted process of internationalization is not the substantial economic transformation that has been occurring since the 1960s but the possibly more far-reaching transformation of states and state policies that has occurred since that time—a process I will illustrate in the chapters on Thailand that follow.

The internationalization of the state, it is worth reiterating is not in itself a merely political phenomenon any more than it is a merely reflexive response to economic forces. Indeed, rather than presenting dualisms such as politics/economics or foreign/domestic as identifying separate realms of activity that compete with one another for primacy, the perspective offered here sees classes and other collectivities as working simultaneously through both foreign *and* domestic arenas, politics *and* economics. The primacy accorded to one or another of the poles in these dualisms, then, reflects not the dominance of one realm over the other but rather the balance of forces in class (-relevant) struggles that involve different fractions of capital with differing geographical ranges.

Such a conception of state power, one differing from that put forward by either neo-liberals or neo-Weberians, has consequences for interpretation of development processes in places such as East Asia. Neo-Weberians have argued that the success of the Northeast Asian NICs is attributable to strong, developmental states taking effective action to maximize the national benefits from participation in international capitalist development (Amsden 1989; 1990; Wade 1990; Evans 1995). The developmental state literature that has emerged from neo-Weberian analyses of Northeast Asian states defines the ideal-type of the interventionist state as neither socialist ('plan-irrational'), nor free market (without planning), but 'plan-rational'—with Chalmers Johnson's influential characterization of Japan providing the most concrete model (Woo-Cumings 1999: 1–2; cf. Johnson 1982).

While their recognition of the importance of state policies that do not merely conform to market forces makes neo-Weberian arguments superior to the strained neo-liberal analyses of East Asian growth put forward by the World Bank and others (World Bank 1993; cf. Wade 1996*b*), neo-Weberian positions are not without limitations. Indeed, neo-Weberians run a serious danger of misspecifying the contributing factors behind East Asian economic growth, and an analysis of the internationalization of the Thai state suggests that factors significantly different from those highlighted by neo-Weberians have contributed substantially to Thailand's rapid industrial transformation.

In concluding the present theoretical discussion, I highlight some of the limits of the developmental state paradigm as an approach to the Thai case by briefly examining Adrian Leftwich's attempt (1995) to systematize various neo-Weberian observations regarding developmental states. Leftwich follows the World Bank in picking out the East Asian NICs, including Thailand, as the success stories to be explained—though he also includes China and Botswana in the analysis—and specifies six features which he suggests are shared to a greater or lesser extent by all of these successful states: (1) a determined developmental elite; (2) relative autonomy of the state; (3) a powerful, competent and insulated economic bureaucracy; (4) a weak and subordinated civil society; (5) the effective management of non-state economic interests; and (6) repression, legitimacy, and performance (Leftwich 1995: 405).

This attempt to distil basic themes from the developmental state literature is problematic in a number of ways, as I illustrate by previewing here aspects of the discussion of Thai development that is undertaken in Chapters 2–4 below. The Thai case is especially noteworthy because Thailand has gone through a very significant industrial transformation in the post-World War II period and was the most rapidly growing of all the Asian NICs from the mid–1980s into the mid–1990s, yet it does not fit most of Leftwich's characterizations of a developmental state. For example, in specifying what it means to have a determined developmental elite Leftwich emphasizes the importance of officials who limit debilitating levels and forms of corruption, yet he immediately acknowledges that Thailand and Indonesia represent cases where there have been instances of spectacular high-level corruption. Leftwich attempts to finesse this by simply reasserting that in comparison to what exists in many developing societies, corruption in the developmental states has been less pervasive—an assertion that carries little weight, given his own admission of the difficulties in defining or measuring corruption and the lack of any data to substantiate the case for the developmentalist NICs (Leftwich 1995: 407). Leftwich also acknowledges that Thailand poses an exception to the claim that developmental states exhibit substantial autonomy (Leftwich 1995: 408), a point which will be illustrated in the historical discussion of Thailand below, where I argue that the Thai state has been integrally connected to the most internationalized fractions of capital.

These do not exhaust the problems. Leftwich also notes that Indonesia, Thailand, and China appear to have had less bureaucratic efficacy and unity than the leading developmental states (Leftwich 1995: 413)—and indeed the subsequent chapters will suggest that unity in Thailand was not great except where elites concurred over policies in their own collective class interests, such as the repression of peasant and labour organizations. Since the Thai bureaucracy is thus acknowledged by Leftwich to have been relatively inefficacious, it would seem to matter little exactly how one evaluates its competence (Leftwich

1995: 414)—a matter that would in any event raise contention (see e.g. Unger 1998). Leftwich's assertion that developmental states have generally had weak civil societies (Leftwich 1995: 415–16) may somewhat accurately describe Thailand throughout much of the post-war period, at least if civil society is used to refer to non-bureaucratic forces besides those of private capital. Yet even here the historical chapters below will illustrate continuing rounds of social struggle emanating from Thai civil society that cannot be overlooked in analysing the development of the Thai state. Moreover, as will be shown, Leftwich's insistence that developmental states have been able to use distribution of the benefits of growth to maintain legitimacy while exercising repression where needed is not particularly helpful in analysing how the Thai state has subdued this civil society (Leftwich 1995: 419). There can be no doubt that all states use some combination of repression and legitimacy-buying distributional strategies, but given the increasingly skewed distribution of wealth and income within Thailand over the recent decades of rapid economic growth and industrial transformation, it is not clear that distributional strategies can be seen as a major characteristic of the Thai state's approach to buying legitimacy.

In short, then, Thailand does not seem to fit the developmental state mould—at least as this mould is articulated by Leftwich. Moreover, it is not only in Leftwich's work that one can find such neo-Weberian attempts to stretch the developmental state model to an uncomfortable extent in order to make it fit the Thai case. Alice Amsden (1995: 796), for example, cites Thai Board of Investment (BoI) contracts specifying what conditions recipients of promotional privileges have to fulfil as an example of how the Thai state, like those of Northeast Asia, disciplined capital (cf. Unger 1998: 75). This is an extremely thin reed on which to hang such a characterization of the Thai state, and the ensuing chapters will strongly suggest that the Thai state was in fact far less disciplinary towards capital than it was facilitating. Indeed, it will be suggested that BoI investment promotion policies are for more representative of the internationalization of the state than of attempts to discipline capital.

It could be the case, of course, that all of this misfit between the Thai state and the developmental state paradigm simply owes to the fact that Thailand is entirely unique. Even if this were so, it would be an important issue for neo-Weberians to address, since theories of economic growth should be able to account for why certain exceptions to general patterns might occur. But it is not entirely clear how exceptional Thailand is, and it is also not clear that the neo-Weberian approach as developed so far is entirely adequate for capturing all the features of the late industrializers that have contributed most to their success. Indeed, recent reassessments of the neo-Weberian literature on South Korea have suggested that the original formulations of the developmental state paradigm overestimated the ability of the state to govern capital (Lim 1998), underestimated both the importance of Japanese investment (Chibber 1999) and the internationalization of Japanese state practices (Woo-Cumings 1995; Kohli 1999), and neglected the unique global geopolitical context of

South Korean growth (Cumings 1999*b*; Pempel 1999). Other accounts have also criticized neo-Weberians generally for underestimating the importance of labour repression and overestimating the political stability and sustainability of South Korean state practices (Bello and Rosenfeld 1990; Hart-Landsberg 1993; Hart-Landsberg and Burkett 2001).

Moreover, while neo-Weberians have done an impressive job of showing that the state played a central role in Northeast Asian economic growth, they have sometimes slipped from arguing this point into making the claim of a region-wide set of similarities in state policies that purportedly account for superior regional performance (Amsden 1995; Leftwich 1995). The latter argument is far less persuasive than the former, and indeed it does not follow at all from the fact that the states played substantive roles in economic development within a given region that the state's role per se helps explain high gross domestic product (GDP) levels throughout that region. Indeed, in Chapter 5 I will argue that a more persuasive explanation of shared high levels of GDP growth throughout East Asia would focus at least in part on a number of socio-spatial factors that are related to the internationalization of states in the region: position in relationship to overseas Chinese networks of capital, Japanese capital, and the US national security state (Bernard and Ravenhill 1995). The particularities of state activities certainly contribute something to regional growth dynamics, but given the variations between activities of states *within* the region it is even more likely that such particularities might say something about differences in the patterns of development *among* the East Asian NICs (Jomo 1997; Booth 1999).

What this discussion calls into question, then, is the issue of whether or not the strong, developmental state identified by neo-Weberians is the only potential progenitor of late capitalist industrialization. The Thai case, I will argue, suggests that weaker states can also facilitate rapid growth and industrial transformation, and that one key to their capability in this regard lies neither in their autonomy (which is often limited), nor in their embeddedness in domestic social structures (which is only one aspect of their embeddeness), nor in their adaptability (which is still being put to the test), but rather in their ability to function as internationalizing class agents, which foment development not so much to produce effective accumulation on a national scale as to enhance possibilities for accumulation by members of transnational hegemonic blocs.

The consequences of this interpretation of state power for issues of the equity and sustainability of Thai growth are obviously significant, and I will elaborate these in Chapters 5 and 6. As an advanced warning, though, I should note one salutary reminder about the limits of any such interpretive project. In her discussion of the developmental state paradigm, Meredith Woo-Cumings notes that Chalmers Johnson's ground-breaking work on the developmental state in Japan was not fundamentally 'an analytic account in search of causal arrows' but rather 'a historical account in search of meaning . . . behind the actions of Japan's policymakers' (Woo-Cumings 1999: 2). Though the account of Thai state policies presented here supports an internationalization of the state approach more than the developmental state paradigm, it too is more a historical account in search of the meanings behind policymakers' actions than a narrow search for causal arrows. To be sure, I will argue that the internationalization processes discussed contribute discernibly to outcomes, and thus act as causal forces. They are not, however, the only causal forces at work in the 'open systems' I analyse. As such, my account of the internationalization of the Thai state is not an attempt to determine the specific causal weight of the different factors producing uneven development and industrial transformation but rather an attempt to foreground contributing factors that have been ignored or misconstrued by analyses operating from within the 'territorial trap'. In short, like Johnson's account of the Japanese state, this account of the Thai state is designed as much as anything else to alter slightly where we look and what we see when analysing the roles played by states in the Janus-faced process called development.

2

Internationalization of the State under US Hegemony: Building the Cold War Regime and Capturing Peasants, 1945–1975

2.1 Introduction

A strong argument can be made that the Thai state has been highly internationalized for a very long time. This is in part a function of the breadth and depth of regional trade and migration networks in the pre-European colonial period, many of which integrally involved Thai dynasties (Reid 1988; 1993). David Wyatt argues that the Ayutthaya-based monarchies had a distinctly cosmopolitan flavour, incorporating substantial numbers of governmental representatives from a variety of Asian trading partners, and representation of foreign interests within the royal court continued into the Bangkok period as well (Wyatt 1984). Hans-Dieter Evers has gone so far as to declare the Bangkok-based Siamese dynasty founded in the late eighteenth century to be a fundamentally trade-based regime (Evers 1987). The importance of trade, in turn, strengthened the importance of foreign merchants and advisers within the royal court.

The incursions of the British and other European powers into nineteenth-century Southeast Asia contributed to further internationalization of the Thai state, particularly after the signing of the Bowring Treaty in 1855. The reforms of the Thai state launched in this context led to not only a dramatic overhaul of the existing bureaucracy but to the employment of enormous numbers of European advisers. William Siffin, for example, notes that during Rama V's reign (1868–1910) there were a total of 549 foreign officials who served in the Thai government, most of these serving after 1880.[1] In 1909 alone, there were some 319 foreigners serving in the Thai government, including 6 general advisers, the general financial agent of the government, 21 legal advisers and

[1] During Rama IV's reign (1851–68) there had been 84 Europeans serving in the government (Siffin 1966: 50).

assistants, 13 director-generals of departments or equivalent, 23 assistant director-generals, and 69 persons engaged in administrative work just below the level of the departmental management (Siffin 1966: 96–7). The role of these advisers in influencing Thai state policies and in representing the interests of metropolitan capital within Thailand has been extensively discussed elsewhere and need not be addressed here.[2] Suffice it to say that their presence manifests the internationalization of capital, while their successful accommodation by Thai elites—who were able to collaborate with them in projects of mutual benefit—speaks to the ways in which the state became internationalized around shared transnational interests (Suehiro 1989; Chaiyan 1994; Thongchai 1994).

Yet the evolution of internationalization processes in Thailand—both within and outside the Thai state—was much more rapid and intensive in the post-World War II period, under the impact of US imperialism, than it had been during any preceding period. The internationalization of capital and the state during the period of British hegemony (roughly 1855 to 1926) had created an administrative framework within which capital accumulation would function after World War II, including the land and labour markets necessary for capitalist accumulation (Chaiyan 1994; Pasuk and Baker 1995). Yet most Thai rural producers remained relatively loosely connected to formally capitalist commodity and labour markets—especially in peripheral regions of the national territory such as North and Northeast Thailand (Anan 1984; Chairat 1988). The urban labour force remained a very small portion of the total national labour force, and was composed primarily of Sino-Thais (Skinner 1957; Hewison and Brown 1994).

The intensification of capital accumulation, triggered by US interventions in Northeast and Southeast Asia, forced a much more rapid development of capitalist market forces, with broad consequences for Thai society. US intervention and the activities of the Thai state in this context moved well beyond the more narrow concerns of British imperialism with natural resource extraction and maintenance of a favourable political climate to embrace numerous aspects of social life, helping usher in a fully-fledged process of industrialization in Thailand.

This chapter outlines the role of the internationalized state in the agrarian and industrial transformation of Thailand during the period of US hegemony (1945–75). I will focus particularly on the way military and police force has been used to buttress primitive accumulation and social transformation in Thai rural society while, in the same process, creating a particular kind of urban labour force.[3] In the first part of the chapter, I illustrate processes of

[2] See e.g. Vella (1955); Holm (1977); Anderson (1978); Brown (1978; 1988; 1992); Feeny (1982).

[3] I use the term 'primitive accumulation', following Marx (1967: 713–74), to indicate the 'original' act of accumulation by capitalists, based on expropriation of the resources and small-scale property held by 'pre-capitalist' agriculturalists and the consequent creation of a landless agrarian workforce available for labour in capitalist industries.

transnational collaboration around development of these institutions at the national level; in the second part of the chapter I note the particular evolution of these processes in Northern Thailand.

2.2 Overview of Thai bureaucratic capitalism and political economy, 1945–1980

The last years of World War II temporarily compromised the position of Phibun Songkhram's ruling military faction of the People's Party, which had collaborated with the Japanese during the war. This provided Pridi Banomyong's civilian faction of the People's Party, which had resisted occupation, a brief opportunity (1944–7) to develop a functioning polity. But a military coup in 1947 placed militarists back in a position of dominance that was not to be relinquished again until 1973, and then only temporarily.

It is during this 1947–73 period that many of the most obvious features of what has come to be called 'Thai bureaucratic capitalism' emerged. Thai bureaucratic capitalism represented a quid pro quo between Sino-Thai capitalists and Thai military bureaucrats. In exchange for state protection and favouritism, Sino-Thai capitalists provided lucrative opportunities for various state bureaucrats, including (in the quintessential form of this relationship) seats on the boards of directors of their companies (Riggs 1966; Sungsidh 1983).

Given the fact that Sino-Thai capitalists were already integrated into various regional production networks, the alliances formed in this process were always already somewhat internationalized, and in this sense the often-noted 'nationalism' of the Phibun regime was more a tool for ensuring opportunities for ethnically Thai bureaucrats than a tool of national economic autarchy. Indeed, as Danny Unger argues, the regional basis of Sino-Thai production networks acted as a deterrent to the Thai state adopting the kinds of fully-fledged import substitution policies that were prominent in other developing countries after World War II (Unger 1998: 55). At the same time, it should be noted that the fact of import-substitution being somewhat limited in Thailand during the 1950s does not mean that protection of domestic markets was unimportant. Indeed, the bureaucratic capitalist alliance constructed lucrative opportunities for domestic investment—including, for example, the development of a protected domestic banking sector that was quickly dominated by leading Sino-Thai families (Hewison 1989*a*). In this sense, the bureaucratic capitalist alliance favoured both 'local' capitalists—such as those ethnically Thai elites who came to control prominent local manufacturing concerns—and those among the more internationalized Sino-Thai who could function effectively under a regime that restricted entrance into certain domestic markets.

Thai bureaucratic capitalism was also internationalized in ways that were extra-economic, however, and the alliance took its specific form within an international hegemonic bloc that evolved in the context of the Cold War. This can be seen in the connections of leading members of the alliance. Phibun maintained power from 1947 to 1957, in part by assiduously playing two competing cliques of bureaucratic and capitalist elites off against one another. These two cliques—the Soi Rajkhru group, of which Police General Phao Sriyanond and Field Marshal Phin Choonhavan were key members, and the Sisao Deves clique, in which General Sarit Thanarat was central—provide the best examples of state–capitalist symbiosis (with international backing) during the period of Thai bureaucratic capitalism. The Soi Rajkhru group brought together a number of leading business people, including key members of the Sino-Thai business community, as well as police and military figures like Phao and Phin. Among the most important members of the business community represented within the group was Chin Sophonpanich, a co-founder of the Bangkok Bank in 1944 (Hewison 1989*a*: 192). Perhaps of equal significance, the Soi Rajkhru group was linked through Phao to the US Central Intelligence Agency (CIA), which supported the Thai police as its favoured force for anti-communist activity (Darling 1965: 113–16; Lobe 1977: 19–27). The Sisao Deves clique has had an enduring impact on Thai development not only because of Sarit's ascendancy to leadership of the Thai state in 1958 (after Phibun was ousted in a 1957 coup), but because of the prominence of other members such as General Thanom Kittikachorn, who succeeded Sarit to power after his death in 1963, and Major-General Kris Siwara, who was to help temporarily push Thanom from power during the popular uprising of 1973. Moreover, Sarit had backing from the US military, and also gained the cooperation of the CIA after Phao had been pushed out of the country in 1957 (Darling 1965: 116–17; Fineman 1997: 247–50).

When Sarit took power, he not only implemented tough anti-communist measures—ushering in what Thak Chaloemtiarana calls a 'politics of despotic paternalism'—but also actively cooperated with the plans of international capitalists for moving Thailand gradually away from state-led economic development (Thak 1979). The most substantial indication of this cooperation is Sarit's response to a World Bank mission to Thailand during 1957–8. Sarit basically followed the bank mission's recommendations to the letter, implementing an industrialization programme that somewhat deepened existing import-substitution measures, while setting up major development agencies desired by the bank such as the Board of Investment and the National Economic Development Board, now the National Economic and Social Development Board (Grit 1982: 114–24; Muscat 1994: 88–92). Under this regime, the importance of the older form of Thai bureaucratic capitalism began to fade, but it was replaced by an only modestly different alliance of interest between Sino-Thai capitalists, Thai state bureaucrats, and international investors in leading industrial sectors, the latter often investing under the umbrella of broad Cold War infrastructure development projects subsidized by US aid.

By the time Sarit died in 1963, US involvement in Thailand was already substantial, and it was to become more so over the course of the 1960s and 1970s because of the use of Thailand as a US military base during the Vietnam War. While only a crude indicator, some sense of the scope of US investment can by gained by looking at the total figures for all forms of aid to Thailand, which amounted to over US$1 billion in economic and US$2 billion in military assistance between 1950 and 1988, with US$2 billion of this total flowing into Thailand between 1965 and 1975 alone (Muscat 1990: 296, 329; Girling 1981: 96). The total amount of regular US military assistance to the Thai armed forces during 1951–71 amounted to 59 per cent of the total Thai military budget; and total US assistance during 1965–75 was equivalent to all of Thailand's foreign reserves (Girling 1981: 96, 236). US aid to Thailand thus made a significant difference in Thai economic performance, contributing to the much lauded gross domestic product growth rates averaging 8.1 per cent per year for 1961–6, 7.5 per cent per year for 1967–71, 6.2 per cent per year for 1972–6, and 7.1 per cent per year during 1977–81 (Warr 1993: 29–32).

US aid and support were of particular importance to the development of two conditions for rapid capital accumulation on a national scale. First, US aid heavily targeted infrastructure development, including roads into rural areas of the country (Muscat 1990: 94–107). Second, US support for authoritarian development under figures like Sarit helped strengthen a repressive state apparatus capable of maintaining effective control over labour and peasant organizations. Aside from assisting in the development of Thailand's infrastructure and in the subjugation of labour, the United States also helped shape the Thai development bureaucracy quite substantially through its training of personnel (Muscat 1990), building an orientation in the Thai bureaucracy that facilitated cooperation with international development agencies and leading players within the interstate system.

The development alliances created in the period of US hegemony have been summarized by Suehiro Akira as a tripod formation, with one leg of the tripod comprised of state enterprises and firms in which the military was involved, a second leg comprised of Thai financial conglomerates and business groups (primarily Sino-Thai), and the third leg comprised of transnational corporations. The three legs of the tripod were connected through joint ventures, and the entire structure was mediated by the military dictatorship and its US backers (Suehiro 1989: 282). I now examine the historical processes of social struggle and alliance building that led to the development of this internationalized tripod structure.

2.3 US Fordism and liberal internationalist imperialism in Asia

US foreign policy in Asia was spearheaded after World War II by a liberal internationalist alliance centred in the Democratic Party and having strong adherents among the most internationalist, most capital-intensive firms (Ferguson 1984; Cumings 1990). This alliance promoted Fordism at home and a liberal variant of imperialism abroad. While liberal internationalists remained at the forefront of US policy-making throughout much of the post-war period up to the ascendancy of the Reagan administration in 1980, they rarely had a clear field and were regularly forced to compromise with the more conservative expansionists centred in the Republican Party, moving in the direction of 'containment', for example, as McCarthyism took hold in the late 1940s (Freeland 1985).[4] Indeed, during periods of the 1950s—especially at the height of the Korean War in 1953, and after the Vietnamese victory over the French at Dien Bien Phu in 1954—the more conservative forces predominated, leading to tactical changes in policy, described under headings such as the shift from 'containment' to 'rollback' (Cumings 1990).

Nonetheless, to understand US activities in Asia, we need to start by briefly examining the major concerns of liberal internationalist US foreign policy planners in the post-war period.[5] These concerns were focused overwhelmingly on the reconstruction of Europe, which was to provide the major market outlet for the US surplus and which was also to be one of the major global engines of industrial capitalist recovery (Kolko and Kolko 1972; Hogan 1987; Leffler 1992). Asia was of lesser significance, but it was not unimportant because Japan's rehabilitation was seen as necessary for reconstruction of global capitalism (Borden 1984; Schaller 1985). Third World countries were largely seen as important in their function as suppliers of raw materials for re-industrialization, and for Asian developing countries this meant that those which were European colonies would be expected to supply resources for the European colonial powers while those which were former Japanese colonies would be expected to contribute to Japanese rehabilitation (Cumings 1981; 1990; Rotter 1987; Kolko 1988).

These broad plans for Asian regional integration, which had evolved by the end of the war (Shoup and Minter 1977), could not be implemented in the desired fashion, however. First of all, there was much resistance to liberal internationalism from the more conservative, expansionist bloc in the US

[4] For discussion of the differences between the liberal internationalists and the conservative expansionists, see Cumings (1990: 23–31).

[5] By US foreign policy planners I mean the leading figures from the executive, the National Security Council, the CIA, the State Department, the Defense Department, and, on occasion, other departments that devote their attention to shaping major contours of the US nation-state's activities in the international arena.

Congress, making it necessary to package economic programmes as military defence (Kolko and Kolko 1972; Cumings 1981; 1990). Secondly, plans for regional integration in Asia were challenged by the uprisings of peasants, linked to communist and nationalist political movements, which welled up in the aftermath of the war—particularly in China, Korea, and Vietnam (Cumings 1981; Kolko 1985; 1988). Rather than being able to implement liberal internationalist designs for economic integration on a tabula rasa, US planners were forced to conceive of mechanisms for subduing unruly peasantries and fitting them into the prescribed model of development. For this reason, too, the deployment of military force came to be a prerequisite for successful integration of the global capitalist economy under US hegemony. In this context, the politics of anti-communism became indistinguishable from the promotion of global capitalism under US military tutelage.

Because of both broad resistance to colonial capitalism and imperialism throughout Asia and the broad ambitions of US foreign policy, the anti-communist project was to be a much more comprehensive and complex undertaking than had been the earlier British imperial efforts. First of all, US foreign policy experts and their allies developed a wide range of approaches to managing local rebellions, including psychological warfare tactics and other counter-insurgency measures, precursors of contemporary 'low-intensity conflict' approaches (Klare and Kornbluh 1988; McClintock 1992). These were employed extensively to combat rebellions in the Philippines and Malaysia during the 1950s, and their success in undermining rural rebellions in these countries made US and British planners confident they could be effectively deployed in countries such as Vietnam (Thompson 1966; 1968; Lansdale 1972; Kolko 1988). In addition, US planners hoped to develop a more complex set of capitalist social relations in the Third World (Rostow 1960), and realizing that this required 'modernization' of a large number of social arenas (Parsons 1966), they set out not just to transform the economy in the narrow sense but rather to transform entire sets of social institutions (Bell 1978; Escobar 1995).

Thailand's process of development evolved in the context of its role in this broad US geopolitical and economic project within Asia. At the most fundamental level, Thailand was initially seen by US planners as a country with the capacity to export rice to rice-deficit areas of Asia (Neher 1980: 170–2). However, the presence within Thailand of a small but important group of entrepreneurs ensured that US economic policies towards Thailand were not driven exclusively by regional rice markets (Neher 1980: 38–62). On the other hand, Thailand's significance to the US foreign policy agenda appreciated substantially as it came to be seen as a regional bastion of anti-communism, a base for US military operations throughout Southeast Asia, and potentially a model in its own right of capitalist development. In the next section, I examine how these interrelated US projects evolved in relation to the development of the Thai military and police, and how the agendas of the US imperial state under Fordism interpenetrated the agendas of the Thai state as it intensified the process of peripheral capitalist development and primitive accumulation.

2.4 US imperialism and the Thai state in the era of US hegemony

Perhaps the most prominent feature of US-Thai interactions in the period of US hegemony was their persistent anti-communist orientation. Yet the Thai state should not be seen as promoting an anti-communist agenda simply out of deference to US policy objectives. The first anti-communist law in Thailand was passed in 1927 (Bowie 1997: 61–2), and though US and British influence in Thailand was already felt by that time there is little reason to doubt that the emergence of strong anti-communist sentiments prior to the Cold War reflected the deep hostility of Thai elites themselves to any radically redistributive agenda, including any such agenda emanating from the provinces and making claims on the wealth that was being siphoned from the provinces to Bangkok. Indeed, the economic plan forwarded by Pridi in 1933, which called for state intervention to prop up agrarian incomes, was opposed by both Royalists and conservative members of the People's Party on the grounds that it was communist (indeed 'Bolshevik') in inspiration (Pridi 2000). While Pridi was eventually cleared of charges of being a communist and allowed to re-enter the government, the episode showed that the elites could readily use charges of communism against movements that might whittle away their class privilege, even before the rhetoric of the Cold War had made opportunistic invocation of the communist threat fashionable.[6]

In spite of the 'local' class basis for anti-communism, however, the use of anti-communist rhetoric by the Thai military and police was a phenomenon that developed much more extensively over the post-war period—and primarily in relation to 'external' factors. When the military coup of November 1947 ousted Pridi's civilian political bloc from state power, the official excuse used was not that Pridi was a communist but that his regime was corrupt and had allowed general chaos to descend upon the economy—inflation having raised prices to eight times their immediate post-war level (Darling, 1965: 57). Conservatives who were opposed to Pridi's liberal regime also implied that he was at least indirectly responsible for the death of the young King Ananda (Rama VIII), for whom he had acted as regent (Darling 1965: 53–5). Thus, on the Thai side, there was no immediate use of anti-communism as legitimation for the return to military rule.

[6] Bowie (1997: 61–2) notes that King Chulalongkorn is reported to have expressed concern over the threat of communism as early as 1881.

Nor do early US reports from Thailand suggest the need for any concern about communism in Thailand. A 1945 report by the Office of Strategic Services (the predecessor of the CIA), stated US interests in Thailand simply in terms of 'maintaining in Siam an open door for American commerce and political influence' (cited in Neher 1980: 2). Two years later, a study of Thailand conducted by the State–War–Navy Coordinating Committee (SWNCC) still indicated that maintaining friendly economic and political relations was the major—and readily achieved—priority. The 29 May 1947 draft of the SWNCC's study insisted that as of that time, 'Communism is not a problem in Siam,' having 'little appeal in a country whose economy is based upon small peasant landowners'. Furthermore, members of the Thai Communist Party were Chinese residents and focused on politics in China rather than in Thailand. Finally, 'Right-wing groups, particularly the army and the police, have far more influence in Siamese politics than left-wing ones,' making the threat of communism even more remote.[7] Thus, SWNCC recommended that to meet the US policy objective of maintaining 'an independent, democratic Siam friendly to the United States, having a stable government and a prosperous economy', no large programme of aid was necessary.[8]

The US plans that developed in Thailand from this point need to be assessed in the context of broader planning for incorporation of Asia into a US-led regional bloc. The central US policy planning document for Asia drawn up by the National Security Council (NSC), NSC 48, expressed fairly clearly the general objectives of policy against which US anti-communism was constructed.[9] Acknowledging candidly that US planners 'should plot our course and make our decisions primarily on the basis of what is good for the United States', the 31 August 1949 draft of NSC 48 goes on to outline an agenda it asserts will be mutually beneficial for both Asians and Americans.[10]

This agenda included implementation of a set of economic principles that are 'widely recognized and accepted, if not always observed'. The principles are as follows:

1. The economic life of the modern world is geared to expansion. . . .

2. Trade between countries based upon government financial aid, special privileges or special restrictions . . . does not contain the elements essential to stability and to enduring survival. . . .

3. Production and trade which truly reflect comparative advantage offer the greatest prospect for general improvement in the standard of living of all peoples and the surest basis for enduring peaceful trade among nations. The

[7] SWNCC, 'SWNCC Second Phase Study on Siam', first draft, 29 May 1947: 1, US National Archives, Division of Research for the Far East, 1946–52, Box 2.

[8] Ibid. 5–6.

[9] NSC 48 was in many ways a companion document to the later and broader National Security Council study, NSC 68. On NSC 48, see Cumings (1990: 157–77). On NSC 68 and its context, see Kolko and Kolko (1972: 470–6); Borden (1984: 18–60); Sanders (1983: 23–50); Cumings (1990: 177–81).

[10] 'NSC 48', Economics Section, 31 Aug. 1949 draft: 1, US National Archives, Judicial, Fiscal, Social Branch, Records of the National Security Council, NSC 48 File.

international trade of countries which the United States is in a position to influence should be permitted, unless overridden by imperative security interests, to reflect the natural and long-established advantages of the several countries in the production of specific commodities.

4. The phenomenal economic progress in the United States, accomplished with the aid of capital and modern technology, has led to extravagant claims by other countries that reliance may be placed largely upon general industrialization for the improvement of the economic condition of their people. This view is no less misleading in Asia than in many other parts of the world. Nearly all the countries of Asia possess special resources and other attributes (some in relative abundance) which can be developed to their particular advantage, but none of them alone has adequate resources as a base for general industrialization. (India, China, and Japan are the only countries which to any degree approximate that condition.) General industrialization in individual countries could be achieved only at a high cost as a result of sacrificing production in fields of comparative advantage. The greatest promise of economic progress, apart from general improvement in agricultural methods and transportation, lies in a highly selective economic development accompanied by a vigorous commerce among the several countries and with other parts of the world.

5. In certain parts of the world, such as Southeast Asia (also parts of South America and Africa), which are natural sources of supply of strategic commodities and other basic materials, the United States has special facilities for assisting economic recovery through its capability of being a large and very welcome customer. . . .

6. In trade with countries under Soviet control or domination, the above principles are not applicable and specific measures to influence such trade to prevent it from augmenting the political power or the capability of such countries for supporting military or other measures against the United States must be taken. . . .[11]

These fundamental principles indicate quite clearly the sort of role US policy-makers had in mind for raw materials exporting countries such as Thailand in the post-war world. While 'soundly conceived manufacturing and other industrial expansion in India and Japan and enlarged trade with other countries of Asia will tend both to improve general economic conditions in that part of the world . . . and to improve markets for United States exports . . . ,' other countries in the region were to concentrate on their comparative advantage as primary commodity exporters. Later drafts of NSC 48 spelled out in more detail the kinds of roles each of the players in this alliance might be expected to play: the pattern of trade 'would involve the export from the United States to Japan of such commodities as cotton, wheat, coal, and possibly specialized industrial machinery; the export from Japan of such items as low-cost agricul-

[11] 'NSC 48', Economics Section, 31 Aug. 1949 draft: 1–4, US National Archives, Judicial, Fiscal, Social Branch, Records of the National Security Council, NSC 48 File.

tural and transportation equipment, textiles and shipping services to Southeast Asia; and the export from the latter area of tin, manganese, rubber, hard fibers and possibly lead and zinc to the United States.'[12]

Such development policies were seen as contributing also to political stability, a matter of great concern to US planners. As the opening to the 31 August draft of the NSC report stated the matter:

> In modern times all of Asia has experienced various degrees and forms of political or social inferiority at the hands of 'western' powers. The two outstanding examples are colonialism and extraterritoriality. Practically all Asiatic peoples suffer more or less from xenophobia often directed especially toward occidentals, but also against one another. There is a wide revival, or in some cases a birth, of nationalism. By tradition and preference Asiatic peoples turn to authoritarian government. In contrast with us, they lack historical experience of liberty and personal experience of individualism. Humanitarianism is almost an unknown sentiment, and the value of the individual is extremely low. They are particularly susceptible to seizure of political power by force or assassination and to the concealed aggression of communism. Internal order and security, old problems in Asia, are exacerbated by struggles for political power, by the aftermath of war, and especially by the arms left in the hands of 'bandits' or 'revolutionaries'.[13]

In this context, as the US planners saw it, the use by the United States of military force to attain its economic goals became both necessary and defensible.

Communism was a target for US imperial policy not so much out of concerns to defend democracy—which, as noted here, was not deemed to be on the cards for most of Asia in any event—but 'because of the cynical attitude of the communists', which would dictate that 'any trade allowed would be to their advantage and would be terminated the moment the communists considered their advantage not sufficiently great'.[14] Thus, the overarching goal of US foreign policy was to expand the sphere of trade and investment controlled by the United States and its pro-capitalist allies, while containing and eventually rolling back the sphere controlled by the Soviets and their communist allies (Cumings 1990; Leffler 1992).[15]

It is in this broad context that the specific development of policies towards Thailand took place. By the late 1940s, Thai rice exports had already been designated as useful to the reconstruction of post-war Asian economies (Neher 1980: 170–2, 362).[16] NSC 48/1, in its final version, had thus called for greater technical assistance to Asian countries under the Point IV

[12] Ibid. Economic Considerations, 7 Oct. 1949 draft, 18–19, NSC 48 File. The document also notes that in 1940 the countries of Asia accounted for 37% of total US imports of basic commodities, including many strategic materials. Thus, re-establishment of Asian countries as raw materials exporters was useful not only for the rebuilding of Asian economic prosperity but for refuelling the economic bases of US politico-military power (p. 17).

[13] Ibid. 31 Aug. 1949 draft: 1.

[14] Ibid. Economic Considerations, 14 Oct. 1949 draft: 21.

[15] On the differences between containment and rollback policies, which were an object of some debate among policy elites, see Cumings (1990: 27–31).

[16] 'NSC 48/1', final draft: 23–4, NSC 48 File.

programme, including loans from the World Bank and the Export–Import Bank to aid agricultural development in Thailand.[17] The culmination of much previous planning came in 1950, however, with visits to Thailand by US Far Eastern Ambassadors for a Conference in February; by the US Economic Survey Mission to Asia, which stopped in Thailand in April; by the Melby Assistance Survey Mission in July; and finally by a high-ranking military officer and a leading CIA official, who came to Bangkok some time during the year to meet with an 'Anti-Communist Committee' organized by the Thai military. As a result of various negotiations that took place around these visits, three major agreements were signed between the US and Thai governments during 1950: an educational exchange agreement under the Fulbright Act, an Economic and Technical Cooperation Agreement, and an Agreement Respecting Military Assistance (Darling 1965: 79–80; Lobe 1977: 19–20; Surachart 1988: 46–7, 66; Fineman 1997: 115, 131–2).

The significance of the military agreement will be examined in some detail below, and the significance of the educational exchange will be discussed in Chapter 3. First, however, I note briefly the significance of the Economic Survey Mission (known generally as the Griffin Mission, after mission leader R. Allen Griffin), which had as its goal 'formulating "a constructive program of aid" to help prevent in Southeast Asia a repetition of the circumstances leading to the fall of China' (cited in Muscat 1990: 20).[18] The Griffin Mission followed earlier economic studies in encouraging the Thai government to promote food and raw material exports (Surachart 1988: 66). It also proposed various technical assistance programmes to Thailand to increase and improve agricultural output.[19] Most interesting for the discussion here, however, is the way in which the Griffin Mission had begun to perceive or portray the relationship between economic development and the threat of communism in Thailand.

The mission's report suggested that Thailand's most significant threat came from 'the large and economically powerful Chinese minority', who control important segments of the economy, 'tend to support whatever government is in power in China', and 'are susceptible to pressure through families and business associations in China now dominated by the Chinese Communists'.[20] The Chinese Communist Party in Thailand, for its part, was seen as small but well

[17] 'NSC 48/1', final draft: 24.

[18] The idea of the survey mission had originally been suggested to US Secretary of State Dean Acheson by Assistant Secretary of State for Far Eastern Affairs, Walton Butterworth, during January 1950 (US Department of State, *Foreign Relations of the United States* (from here on *FRUS*) 1950, vi: 4n.; Fineman 1997: 114–15). The proposal was put forward in the context of attempts to develop a pro-capitalist Southeast Asian regional association.

[19] US Economic Survey Mission, 'Needs for United States Economic and Technical Aid in Thailand: Report No. 4 of the United States Economic Survey Mission to Southeast Asia', May 1950: 6, US National Archives, Records Relating to the Economic Survey Mission to Southeast Asia, Box 1 (hereafter, 'Griffin Mission Report'). Page number references for the Griffin Report are to the National Archives pagination for the report, rather than to the pagination of the original report itself.

[20] 'Griffin Mission Report': 12.

organized and devoted, among other things, to 'welding all Chinese labourers into a strong, disciplined unit which could be used to paralyze the rice-milling and other industries, as well as the country's internal and export trade'. The Party was also committed to 'increasing the flow of propaganda largely directed against the United States, through the medium of controlled Chinese newspapers, pamphlets, and other publications. . . .'[21] While Pridi's faction was not seen as communist, it was regarded as the faction the Communist Party was 'most likely to approach'.[22]

The Griffin Mission recognized that there was 'hardly any important economic urgency in Thailand', but the dangers posed by the communists created a 'political urgency', which could be dealt with through 'A quick gesture calculated to impress government leaders and the people—particularly the educated elite in Bangkok. . . .'[23] Furthermore, Thailand was recognized as a particularly attractive place for the US to make such a showing, because 'The absence of Western colonialism in Thailand has resulted in the absence of the extreme anti-Western passions which inflame much of the Far East, thus making it easier for the Thais to cooperate with the United States politically than it is for some other governments.'[24]

The ideas expressed in the Griffin Mission report highlight nicely the directions in which US–Thai relations had evolved just prior to the beginning of the military build-up in Thailand. Throughout the first five years of the post-war period, Thailand had been regarded as a stable country with few economic problems and no substantial internal threats. Its significance for the US was mainly as a rice and tin exporter, and, more peripherally, as a place where a small number of US investors could hope to reap rewards. Its role in relationship to Asia more generally was seen as that of a food and raw materials supplier and a base of pro-capitalist stability in a sea of nationalist and communist turmoil. As that sea roiled in the late 1940s, Thailand's vaunted political stability and pro-capitalist orientation came to be seen as even more valuable—particularly if it was portrayed as more threatened. Thailand was useful for US regional projects in two ways: first as a base for military activities throughout the region; second, as a potential model of prosperity under capitalist development (*FRUS* 1950, vi: 1530; Neher 1980: 251–2). At home, however, the task of selling the US Congress on overseas economic and military assistance had come to depend on the ability to portray US allies, and thus the US itself, as threatened by communism (Kolko and Kolko 1972). It is therefore unsurprising that US Secretary of State Dean Acheson, in a March 1950 memo to President Truman, justified US military aid to Thailand by invoking an internal Chinese Communist threat (*FRUS* 1950, vi: 42; Fineman 1997: 114).[25] Acheson's claims

[21] Ibid. 13. [22] Ibid. 11–12. [23] Ibid. 91. [24] Ibid. 17.

[25] Acheson's crucial role in selling policies with broad political economic purposes under the emotionally loaded and politically expedient banner of anti-communism is well established in a long line of works on US diplomatic history. See e.g. Kolko and Kolko (1972: 470–5); Sanders (1983: 31); Borden (1984: 43–50); and Cumings (1990: 42–5).

were not backed by reports from the field. For example, in a 6 December 1949 memo to Acheson, US Ambassador to Thailand Edwin Stanton argued that reports by Thai authorities that Thai institutions had been infiltrated by communists were 'exaggerations motivated by [the] desire to urge [the] U.S. to send more and faster aid' (cited in Neher 1980: 218). Nonetheless, the Griffin Mission, which did its brief survey of Thailand one month after Acheson sent his memo to Truman, hewed to the Secretary of State's line.

To be sure, US planners may well have sincerely felt that *if* Sino-Thai communists achieved any power they might threaten Thai stability by undermining Thailand's assigned role as a trade-oriented country, exporting rice to other capitalist nations. If communists seized power in Thailand, they might cooperate with China and pull Thailand out of the 'pro-Western' alliance the United States was attempting to build. More dauntingly, it appeared that the existing Thai government itself was threatening to move in this same direction when, in 1949, it began to talk of not renewing an agreement to sell rice to Japan through the Supreme Command, Allied Powers, so that it could have more latitude to pursue the most favourable markets. As Arlene Neher summarizes the significance of this move (1980: 381):

> Thailand's implied threats to sell rice to China rather than to Japan were among the factors which led the United States to extend more tangible commitments to Thailand beginning in 1950 . . . The United States offered the alliance and the aid to Thailand not only because of the communist threat, as is generally supposed, but also because some tangible commitment to Thailand was necessary in order to persuade Thailand to play the economic role in Asia which America had envisioned for it.

If the existing Thai government might waffle in its commitment to play the US-assigned role, then surely a regime in which the Communist Party played a part would be even more unreliable.

That the Communist Party of Thailand (CPT) was too small and ineffectual at this point to pose any threat was of far less significance to either US or Thai elites than the fact that the assertion of such a threat had become *de rigueur* in the practice of Cold War era development politics. In this context, sorting out precisely what various elites actually believed as opposed to what they opportunistically claimed is unnecessary.[26] Communism and threats to the broad project of regional capitalist development and integration had become fused in the minds of most planners, and thus to argue for capitalist development policies was precisely to argue against communism—and vice versa.

[26] However, for an excellent discussion of the opportunistic manœuvring carried out by Phibun and the US embassy, see Fineman (1997).

2.4.1 The development of the military and the police under US hegemony

As US geopolitical manœuvres evolved in the context of the post-World War II political economy, the potential value of Thailand not only as a site of export rice production but as a military ally came to be recognized on both sides of the alliance. The political transformation was stimulated by two major factors, neither 'internal' to the Thai political economy per se. First, the success of the Chinese revolution in 1949 and the rebelliousness of Korean and Vietnamese peasants in resisting the reimposition of colonial or neo-colonial regimes led US policy-makers to increase their military commitments in Asia, particularly in preparation for the Korean War, but also in relation to the project of controlling Southeast Asia (Kolko 1988; Cumings 1990). Second, and dialectically related to the first factor, the US political environment itself became much more stridently anti-communist by the late 1940s, as the liberal internationalist and more conservative blocs jockeyed for position and vied to outdo one another in asserting the need for military vigilance—the liberal internationalists often using the communist threat as an excuse for military spending that was favoured on essentially economic grounds (Borden 1984: 44–57). By 1950, these factors congealed into a much more forthrightly militarist policy, signalled by the issuance of the now widely noted policy document, NSC 68, and etched more bloodily upon the historical record by the outbreak of the Korean War.

Three years earlier, in April 1947, US State Department Southeast Asian Division Assistant Chief Kenneth Landon had written a general position paper calling for a Mutual Security pact between Thailand and the United States. The paper was sent to SWNCC in July, but was not acted upon for three years, in spite of offers of military cooperation from a Thai representative later in 1947 (Neher 1980: 191–3). The failure to act sprang from a variety of sources, including the previously mentioned SWNCC assessment and lack of US confidence in Thai political stability and anti-communist commitments. These uncertainties were exacerbated somewhat by the coup of 9 November 1947, which the US embassy failed to anticipate and initially miscomprehended—incorrectly thinking it was headed by Phibun when in fact Phibun only joined later (Fineman 1997: 41).[27] When they did begin to sort out events, Landon and Ambassador Edwin Stanton favoured withholding US recognition of the new regime, and the State Department did not finally offer such recognition until early 1948, using the bait of recognition in the interim to manœuvre the Democrat Party and Khuang Aphaiwong into power (Fineman

[27] The major plotters were actually Lt.-Gen. Phin Choonhaven, Col. Phao Sriyanond, Col. Kat Katsongkhram, and Col. Sarit Thanarat (Fineman 1997: 37). Phibun only agreed to join the group after the fact.

1997: 41–3; Neher 1980: 203–4).[28] More importantly, however, failure of the US government to act on military aid proposals for Thailand reflected the limited importance State Department planners accorded Thailand in 1947.

The sense of Thailand's importance was transformed in 1948, however, by the evolution of the Cold War, by increased US commitments to supporting anti-communism in Asia, and by Phibun's increasingly successful attempts to cast his regime as staunchly anti-communist in the favoured US mould (Fineman 1997: 74–84). The evolution of international political struggles and internal US politics by 1948 had made Phibun's past record of collaboration with the Japanese irrelevant to most US planners and the conservative cast and politico-military dominance of the forces he represented far more decisive (Neher 1980: 66–7). As Landon put it, in the interests of stability, the United States had decided that it 'had to go along with those who were in charge of the store' (cited in Neher 1980: 56). Thus, at the end of 1948, Stanton requested military aid for Phibun's regime (Fineman 1997: 84). Yet even at this point, no military aid was immediately forthcoming, with the fight for approval from the State Department continuing through 1949 (Fineman 1997: 90ff.). Not until 1950, in the context of a much expanded US military budget and intensified military activities in Asia under the rubric of 'containment', was the Mutual Security Assistance pact suggested by Landon three years earlier finally signed.

The 1950 agreement set in motion a long period of substantial US military assistance to Thailand, whose quantitative dimensions have already been mentioned. Notably, the sheer quantitative significance of US aid was perhaps greatest in the 1950s, when it amounted to as much as 20 per cent of total Thai government expenditure in some years. Indeed, in 1953, US military assistance was equivalent to 246 per cent of the Thai military budget (Surachart 1988: 195). Moreover, in addition to providing military aid, the US also trained thousands of Thai under the International Military Education and Training Program.

More important than the sheer quantitative power of US military assistance was the qualitative significance of the US-Thai military relationship. US military aid created a context in which US advisers could effectively propagate their own anti-communist values and their particular interpretations of the nature of the communist scourge (Chai-anan, Kusuma, and Suchit 1990: 26–8). Over time, many leading Thai and US military figures became not only allies but personal friends, with shared values and perceptions of the crucial issues confronting them (Darling 1965: 106–7: Fineman 1997: 181). Moreover, the US

[28] Unofficially, however, the US attitude towards the coup group seems to have been tolerant. A 10 Nov. 1947 telegram from Landon to the Secretary of State (just two days after the coup), states: 'We gather from discussion SEA Dept. resigned to acceptance Phibun regime if latter successful in coup . . .' State Department Files, 1945–9, US National Archives, 892.00. Moreover, when Phibun and the coup group ousted Khuang in April of 1948, US recognition was forthcoming within a month (Darling 1965: 64–6; Fineman 1997: 57).

move to pull the Thai military on board the anti-communist bandwagon created a highly consequential fusion of US and Thai state power and deeply affected the character of development politics over subsequent decades. Phibun's position was buttressed, as was the position of officers whose participation in the 1947 coup had helped bring him back into power, including Phao, Sarit, Phin, Thanom, Kris, and Lt.-Colonel Phraphat Charusathien—in short, virtually all of the military figures who were to dominate the highest echelons of the state over the next 25 to 30 years (Surachart 1988: 47–54; Chai-anan, Kusuma, and Suchit 1990: 29–30). By contrast, Pridi's faction was pushed entirely off the political stage. Indeed, between the November 1947 coup and 1949, six of Pridi's allies had been murdered by Phao's forces, and the eventual fully-fledged US backing for Phibun's regime implicitly endorsed this violent suppression of the opposition, which continued sporadically throughout most of Phibun's reign (Darling 1965: 71; Neher 1980: 207).

In the process of throwing its weight behind Phibun, the US gained a strong ally for its politico-military interventions in Southeast Asia. Phibun's regime refused to recognize the Communist government in China and recognized the Kuomintang instead; it sent 4,000 troops to back the US war effort in South Korea during 1950; it became one of the first to recognize the US- and French-backed regime of Bao Dai in Vietnam during 1950; and by 1971 this Thai cooperation with US war aims in Vietnam had evolved to the point that 10,000 Thai troops were sent to fight there (Darling 1965: 78; Surachart 1988: 46; Chai-anan, Kusuma, and Suchit 1990: 97).

The US military aid tap, once turned on, helped the Thai military elites, centred heavily in Bangkok and the Central region of the country, to consolidate power and to provide patronage to various political and economic allies (Riggs 1966: 249–54; Hewison 1997: 3–10). It also strengthened their ability to demand a share of the economic spoils. Thus, aside from shaping the anti-communist character of the Thai military, the US alliance helped consolidate the military's position as a crucial player in the broader political economy.

Since there was in fact no powerful domestic opposition to Phibun's ruling clique once Pridi's faction was suppressed, much of the military politics of the 1950s devolved on struggles within the ruling coalition itself. The struggle between Phao and Sarit to determine who would replace Phibun is the crucial exemplar, and the development and resolution of this struggle indicate the degree of interpenetration of the US and Thai state apparatuses.

Sarit, as noted earlier, had backing from the US military, while Phao gained most of his backing from the CIA. Every member of the ruling coalition was both anti-communist and opportunist, and in the stridently anti-communist environment that had overtaken the US foreign policy establishment during the 1950s, the opportunist moves generated intense scrutiny and much concern about Thailand sliding down the slippery slope towards neutralism and then communism. Of particular concern by the mid-1950s were both Phibun and

Phao's manœuvrings, which some US officials feared might be leading towards normalization of relations with China.[29] In addition, the United States evinced concern that Pridi, now living in exile in China, might be allowed to return, with destabilizing consequences for Thai politics.[30] Furthermore, though US leaders had generally indicated support for Phibun as a means of balancing the competing ambitions of Phao and Sarit, some concern was also expressed regarding Phibun's attempts to liberalize the political atmosphere during 1955–7.[31] Finally, though the CIA had worked closely with Phao, and had reportedly even helped him establish the contacts with Kuomintang opium dealers that substantially increased his wealth and power, there was growing concern among many US officials by the mid-1950s that Phao's well-publicized opium exploits were becoming a political liability and that having Phao succeed Phibun would be undesirable (Flood 1977: 1–2).[32]

It was in this context that Sarit, who had quietly shouldered Phin aside and acquired increasing power within the military, launched a 1957 coup, ousting both Phibun and Phao. Recent research by Daniel Fineman supports the contention that the US had no direct role in this coup. Moreover, many CIA officials certainly disliked Sarit's usurpation of power, given that it resulted in removal of their leading ally, and given that their operations immediately came under attack from Sarit as part of his effort to consolidate control (Fineman 1997: 239–46).[33] Nonetheless, the US military and the guns-first approach it favoured in Southeast Asian conflicts were ascendant within the Eisenhower administration in this post-Dien Bien Phu period, when the US intensified the projects of creating an independent South Vietnamese state and building regional political and military support for the effort (Kolko 1985; Kahin 1986). As such, it is likely that Sarit acted believing that a substantial element in the US foreign policy establishment would eventually approve of—or at least tolerate—the seizure of power by a military leader[34]—and, indeed,

[29] For examples, see *FRUS* 1955–7, xii: 840–1, 854–5, 875–81, 890–1, 893–4, 916–17.

[30] See e.g. ibid. 846–7, 854–5, 910–12, 919–21. A 21 March 1956 despatch from Ambassador Max Bishop to the State Department asserts that Phao began cultivating contacts with Pridi's followers in 1952–3 and shortly after that with the Chinese Communists. US Department of State, Central Files, US National Archives, 792.00/3–2156.

[31] *FRUS* 1955–7, xii: 928–30. The despatch from Bishop, cited in the above note, indicates concern that Phibun's relaxation of press censorship was leading to more media attacks on the United States. See Fineman (1997: 205–42 *passim*).

[32] On Kuomintang opium operations in Northern Thailand, see Race (1974: 99–102); and McCoy (1991).

[33] An earlier analysis by Thadeus Flood argues that US Ambassador William Donovan (a CIA man) and another close friend of Phao's, CIA chief of station John Hart, actually helped engineer Phao's downfall by training anti-Phao police recruits and promising Sarit more heavy equipment (Flood 1977: 2; Lobe 1977: 28).

[34] Immediately after the 17 Sept. coup, Bishop cabled the US Secretary of State John Foster Dulles to indicate that he would inform Sarit of his willingness to meet with him personally 'to hear what he has to say', unless the State Department objected (State Department Files, US National Archives, 792.00, Box 3909, 17 Sept. 1957). On 23 Sept., Dulles telegrammed back to request a delay in the meeting Bishop had scheduled with Sarit on the grounds that: 'We in great danger giving impression we approved or actually helped engineer coup.' Another telegram from Dulles on the same day hit upon a more subtle way of

rapprochement between Sarit and the CIA began rather quickly after the coup (Fineman 1997: 247).

If the US government was uninvolved in the 1957 coup, there is strong but necessarily circumstantial evidence that it supported and even encouraged the subsequent coup that Sarit's forces carried out in late 1958. Despite Sarit's 1957 coup, US policy elites continued to be concerned about the general political climate in Thailand. An 11 October 1957 telegram from US Ambassador Max Bishop to Secretary of State John Foster Dulles argued for the necessity of convincing political leaders in Thailand of the seriousness of the communist threat, including the need for the Southeast Asia Treaty Organization.[35] Bishop stated that 'As initial step toward solving this problem consider it necessary achieve drastic reduction influence of left-wing press and small but vociferous and well-organized pro-Communist groups . . . If Thai leadership responds by initiating action eliminate these subversive influences, we must of course be prepared encourage and support them.'[36]

US leaders had an ideal opportunity to impress these points powerfully and personally upon Sarit when he required medical treatment for cirrhosis of the liver. US leaders arranged for Sarit to receive treatment at Walter Reed Hospital, in Washington, at US expense, which he did from late January of 1958 to the end of June of that year. Memos from various US officials during Sarit's visit emphasized the desirability of driving home for him during his stay in the United States the importance of the anti-communist agenda—and particularly the need to stifle anti-American or left-wing opinion in the Thai media (*FRUS* 1958–60, xv: 977–1038). Towards this end, meetings were arranged with both Dulles and President Eisenhower (Surachart 1988: 77), and Assistant Secretary of State Walter Robertson also visited the convalescing Sarit to lecture him about criticisms of the United States carried by Thai papers, including one owned by Sarit himself (Fineman 1997: 251). In addition, according to a former CIA operative who worked in Thailand in the 1960s, word around the agency at that time was that the CIA leadership had used Sarit's medical stay as an opportunity to patch up its own relationship with him[37]—and the prospects for this were in any event helped by the removal of former Thailand station chief John Hart and the assignment of Robert Jantzen, who visited Sarit regularly in Washington (Fineman 1997: 250).

Sarit made a brief return to Thailand at the end of June, before going to England for further convalescence, but when he returned to Thailand in

accepting Sarit's ascendance: 'Department taking position question recognition does not arise in connection Thai government change in view of fact King remains chief of state' (Department of State Central Files, US National Archives, 792.00, Box 3909, 23 Sept. 1957.)

[35] Thailand had in fact been one of only two Southeast Asian nations to join this organization when it was founded in 1954, but despite its membership the US Embassy was concerned by 1957–8 about local 'anti-American' sentiments that might encourage Thai withdrawal from the organization (Darling 1965: 108, 137, 150–1, 163).

[36] US Department of State, Central Files, US National Archives, 792.00, Box 3909, 11 Oct. 1957.

[37] Author interview, former CIA official, 13 July 1997.

October 1958, many of the aggravations that had stimulated US concern during Phibun's tenure were still festering, including concern that the Thai government was not acting decisively enough against demonstrators and that Pridi might yet be allowed to return.[38] Sarit allayed these concerns with another coup, on 20 October 1958.

The coup gave Sarit's US backers the kind of strictly controlled domestic political environment they desired. Sarit's forces rounded up communists, banned left-wing political parties, implemented martial law, and harshly censored the press, raiding bookshops and schools to seize communist literature and films. At least 134 persons were arrested during the first few days of the coup, including opposition assemblymen, newspaper editors, writers, labour leaders, teachers, students, and businesspeople. By the end of 1958, Sarit's forces had arrested 1,080 communist suspects, most of them socialist and liberal politicians. Twelve newspapers were closed. Twenty-four of the labour leaders arrested were to remain in prison for 5 years, and several others remained in prison for 8 years. Three of the leftist politicians were executed in 1961–2, including well-known Northeastern parliamentarian Khrong Chandawong (Darling 1965: 187–8; Mabry 1979: 48; Thak 1979: 148; Surachart 1988: 86; Tienchai 1993: 160).[39]

Unsurprisingly, given the background to Sarit's moves, the response from Washington was highly positive. The US Ambassador, Alexis Johnson, immediately endorsed Sarit's actions.[40] A US State Department official, when informed of the coup, stated bluntly, 'We are gratified to hear that,' since Sarit was 'most definitely favorable to the United States' (cited in Darling 1965: 188). One year after the coup, Johnson lauded Sarit's performance in a despatch to the Department of State, noting that in spite of the fact that the coup and its aftermath 'unquestionably constitute a setback for the trend, however faint, toward a more democratic form of government', nonetheless,

> We need not . . . feel self-conscious about our support of an authoritarian government in Thailand based almost entirely on military strength . . . [because] aside from the practical matter of Thailand's not being ready for a truly democratic form of government . . . the United States derives political support from the Thai Government to an extent and degree which it would be hard to match elsewhere.

[38] See the telegrams from Ambassador Alexis Johnson to Dulles, 9 Sept. 1958, and from Dulles to the Embassy in Bangkok, 23 Sept. 1958. US Department of State, Central Files, US National Archives, 792.00/9–958, 792.00/9–2358.

[39] See also the telegram from Ambassador Alexis Johnson to Secretary of State Dulles, 28 Oct. 1958, US Department of State Central Files, US National Archives, 792.00/10–2858.

[40] For Johnson's discussion with Sarit's representative on the day of the coup, indicating that no question of recognition would be raised, provided the King did not object to the coup, see the telegram from Johnson to Dulles, 20 Oct. 1958, US Department of State, Central Files, US National Archives, 792.00/10–2058. See also, Surachart (1988: 86).

Furthermore:

Sarit's concepts and actions as we perceive them approach the Department's definition of the 'happy medium' from the standpoint of US interests as a situation which encompasses 'a military regime "civilianized" to the greatest extent possible and headed by a military leader who saw security and development in perspective and thereby evidenced political leadership of the type required in a developing society.'[41]

With this kind of uninhibited US support, Sarit embarked on a long reign of 'despotic paternalism', which has been extensively described elsewhere and need not be detailed here (Thak 1979).

The tenor of the US agenda and its unflagging commitment to strict authoritarianism during the Sarit years is captured by the response of the US consul in Chiang Mai, George M. Barbis, to a seemingly trivial issue, the 1960 publication by a Chiang Mai newspaper called the *Northerners* of an article contrasting the Soviet space effort favourably with the US effort. The paper that printed the article had, according to the consul, been giving excellent coverage to United States Information Service releases and carried a special Fourth of July issue that praised the United States. It also had, however, 'a past of anti-American, leftist-leaning editorializing' and had been closed down during 1959 for printing an allegedly subversive poem. Barbis noted that the paper had been allowed to reopen in 1960 'and gave the impression of having learned its lesson'. The new article, however, rekindled concern, and, Barbis asserted, 'may be an indication of what we might expect in the Chiengmai press should present controls be relaxed'.[42]

What is important in these sorts of manœuvrings is that they show how deeply Thai 'internal' politics had become imbricated in the larger geopolitical manœuvrings of the United States within Southeast Asia—as well as manœuvrings on the US home front. Sarit's 'despotic paternalism' was not a narrowly domestic project to which Washington mutely or opportunistically subscribed but a joint venture that the US state helped cultivate and which Sarit's forces enthusiastically carried forward.

2.4.2 The CIA, the Thai police, and counter-insurgency as development

The police, as noted earlier, were the domain of Phao and his CIA backers. The latter hoped to use the police as a more subtle tool of communist suppression at the local level. US efforts in this area need to be seen in the larger context of the psychological warfare operations the CIA favoured as an element of counter-insurgency practice. In May of 1953, the US National Security Council recommended that C. D. Jackson report to the next council meeting

[41] *FRUS* 1958–60, xv: 1096–9.

[42] Letter from American Consulate, Chiengmai, to Douglas Batson, American Embassy, Bangkok, 7 Sept. 1960, US National Archives, RG 59, Bureau of Far Eastern Economic Affairs, Office of Southeast Asian Affairs, Thailand Files, 1960–3, Box 1 (1960), Chiengmai Consulate.

on possibilities for psychological operations in Thailand (*FRUS* 1952–4, xii: 670). Jackson, considered a leading expert on psychological warfare ('psywar') strategy, carried out similar assessments of a number of countries, and was asked to carry out the assessment of Thailand in the context of the appointment of William 'Wild Bill' Donovan to serve as Ambassador there.[43] Jackson's top-secret study, which was issued later in 1953, is not especially interesting for its insights regarding Thailand, which are minimal (cf. Fineman 1997: 171–2), but is significant for the broad conceptualization of counter-insurgency in Thailand it articulates.[44]

Jackson's report begins by invoking the usual threats of Pridi's possible collaboration with the Chinese Communists and communism among the Sino-Thai (p. 3).[45] Against these threats, 'the proposed new Western strategy, based upon US leadership in Southeast Asia, must include military, economic, political, and psychological programs,' and not only is effort required in all these areas but 'the several efforts must be mutually reinforcing' (p. 4). The report asserts that it is in the political and psychological areas that the communists currently pose the biggest threat and that effective response to this requires giving nationalist aspirations their due. But, the report warns, the US has been dealing with nationalism 'solely through the urbanized, westernized, upper and middle classes' and 'must learn eventually to associate . . . with Asian nationalism, as the communists do, at the village level . . . In particular we must meet the communist military threat at the village level—where it is the most deadly—by developing local resistance forces, and guerrilla bases deep in communist held territory' (pp. 4–5). Success in this project was not merely a matter of relevance to Thailand but would also strengthen the position of anti-communist forces in Malaysia, Vietnam, and elsewhere in Indochina (p. 5). This regional significance made Thailand 'the principal geographical base of the proposed US psychological strategy for Thailand and adjacent areas, and the central concept of this strategy should be the development in Thailand of programmes which gradually create a climate of victory, and in other ways project U. S. leadership, throughout the area' (p. 6). In short, at multiple geographical scales, the 'psywar' strategy for Thailand envisions the extension of US power from current bases of strength into the countryside and into areas of present communist strength, using multifaceted, village-level operations.

The specific measures suggested include, in Phase I, maintaining all current programmes of US–Thai economic and military cooperation, and in Phase II, extending 'paramilitary and other programs beyond the borders of Thailand', working with indigenous leadership wherever possible (pp. 26–7). Among the specific psychological objectives to be attained were to increase among the

[43] Donovan, the former Office of Strategic Services (OSS) officer, had a long history of connections in Thailand (see e.g. Neher 1980: 13ff).

[44] Psychological Studies Board, PSB D-23, 14 Sept. 1953, Declassified Documents Reference System, 1994, 000556–7, WH 120. All page number references in the following paragraphs are to this document.

[45] The report also castigates the press, which is 'characterized by irresponsibility' (p. 20).

Thai 'the belief that U. S. support in Southeast Asia will facilitate the solution of major political and economic problems in the area—including rural development and land reform—in a manner satisfactory both to Asian and Western interests, without any taint of disguised U. S. colonial aspirations' (p. 28). The 'Basic Tasks' necessary to facilitate attainment of the objectives included commitments to rural development and health projects, improvement of strategic roads and railroads, and special aid to the Northeast. Other tasks included development and expansion of overt information programmes, reduction of Thai vulnerabilities to communist subversion along with combating communist influence on opposition politicians, and development and expansion of indigenous guerrilla and paramilitary forces (pp. 28–30).[46]

The significance of the 'psywar' study is that it illustrates how from virtually the outset of US intervention in Thailand, the use of repressive force was seen to be necessarily linked with the need to promote rural development and what later came to be called 'winning hearts and minds'. Thus, the development of rural counter-insurgency capacity within branches of the Thai police backed by the CIA should be seen as part of the broader attempt to transform the Thai countryside so as to integrate the peasantry more fully under pro-capitalist development schemes.

It is from this perspective that we can understand the creation of the various special police units sponsored by the CIA and the Thai government. Beginning with Donovan's tenure as Ambassador, during which he developed an especially close relationship with Phao (Lobe 1977: 25: Fineman 1997: 181), a number of different paramilitary projects were hatched under police control and with funding from the US Office of Public Safety (OPS) and other sources. These included the Police Aerial Reconnaissance (Resupply) Unit (PARU), which specialized in parachuting long-range patrols into isolated areas to collect intelligence and sabotage enemy facilities, and the Border Patrol Police (BPP), which were responsible for defence along Thailand's international frontiers. The CIA helped create these organizations under general directives from the NSC and backed them through the activities of a Miami-based front organization known as the Overseas Southeast Asia Supply Company, or Sea Supply (Lobe 1977: 24; Fineman 1997: 182).

OPS grants during the 1950s were minimal, accounting for less than 1 per cent of the total police budget. However, substantial aid was also given to Phao through other sources, and this enabled him to provision his well-paid forces with their own tanks, artillery, armoured cars, air force, and naval patrol

[46] More specific recommendations for activities are included in the report, but many are censored from the declassified document. Subsequent documents from the Operations Coordination Board indicate that virtually all of the activities recommended in PSB D-23 were carried out, and Daniel Fineman asserts that in fact PSB D-23 become the operative blueprint for US policies towards Thailand through 1954 (Fineman 1997: 170–1). See Memorandum to the Executive Officer, Operations Coordinating Board, 27 Jan. 1954, with attached Annexes, Declassified Documents Reference System, 1995, 002964–6, WH.

vessels (Darling 1965: 114).[47] The effect of US aid to the Thai police in this period was also qualitative and was important for the links it helped establish between personnel and the development of basic counter-insurgency approaches and methods of operation it fostered.

Because of the importance of these personal contacts, Phao's fall from power in 1957 had temporarily negative consequences for Thai police/CIA autonomy. While total CIA and Civil Police Administration/OPS grants actually increased after Sarit came to power, Sarit reduced police personnel and took away police control over heavy and sophisticated armaments, while absorbing PARU into the army for several years and the BPP into the Provincial Police (Lobe 1977: 28: Fineman 1997: 247).

This setback was only temporary, however, and negotiations between Sarit and the CIA quickly resolved differences and re-established CIA operations under Sarit's control (Fineman 1997: 247). Moreover, the position of the police was strengthened again by the 1960s in response to two exogenous factors. First, John F. Kennedy gained executive office in the United States, bringing the liberal internationalist policy elites back into the driver's seat and emphasizing a highly development-centred approach to counter-insurgency, of which the President was highly enamoured (Paterson 1989). Thus, the US Agency for International Development (USAID) began to play an even more prominent role in US foreign policy and became the home to OPS (Lobe 1977: 33). Second, a mid-1960 coup in Laos led by Captain Kong Le undermined US attempts to build a right-wing government there, and in 1961 Kong Le's neutralist army fought along with the Pathet Lao to inflict a defeat on US-backed forces (Devillers 1970: 46–7). The intensification of activity along the Mekong increased fears that insurgency would spread to Northeast Thailand and to Northern hill-tribe areas. Thus, US proposals to upgrade PARU and the BPP for activity in Laos were accepted by Sarit, who felt in any event that these units were now firmly under his control (Lobe 1977: 34–7).

Out of this sequence of events, the OPS was revived and the Thai police and its US advisers began a new round of activities which were to inject the activities of the Thai state deeply into the Thai countryside. Many accounts of this process have been written and need not be reviewed here.[48] What is important is to note both the integration of the Thai and US counter-insurgency institutions and the integration of the various activities considered part of counter-insurgency. Indeed, by the 1960s, the integrative approach had become infectious. On the one hand, whole new development bureaucracies were created to coordinate the wide-ranging activities. The prime example of this

[47] It should be noted that the upper echelons of the Thai police were also well funded through Phao's enormous revenues from opium smuggling and other illegal activities (Darling 1965: 115; Thak 1979: 83–4; Fineman 1997: 144).

[48] The most detailed account of the use of police paramilitary forces in counter-insurgency is Thomas Lobe's monograph (Lobe 1977). Also valuable are Tanham (1974) and Ralph McGehee's account of his work for the CIA in Thailand during the 1960s (McGehee 1983). An important insider's account by a Thai participant is provided by General Saiyud Kerdphol (Saiyud 1985).

was the creation of a Ministry of National Development in 1963 (Siffin 1966: 177). This ministry, which was abolished in 1973, took over many of the functions of the Ministry of Agriculture and Cooperatives and departments such as the Highways Department, within the Ministry of Transportation. On the other hand, the agencies of the repressive state apparatus developed their own multifaceted developmentalist approaches to counter-insurgency. The BPP were encouraged to engage in various forms of 'civic action', such as digging wells, building schools, and helping modernize agriculture (Lobe 1977: 70–87). The military developed Mobile Development Units (MDUs), which built highways and latrines, among other activities (Bell 1978: 64; Muscat 1990: 159–60). Meanwhile, the Ministry of the Interior developed the Office of Accelerated Rural Development, which basically functioned as the flip-side of the counter-insurgency policies developed by the US-backed and trained Communist Suppression Operation Command (CSOC, later Internal Security Operations Command, ISOC) (Saiyud 1985: 38–9).

A particularly noteworthy feature of all of this counter-insurgency inspired developmentalism is that it began long *before* there was any serious Communist Party threat in the rural areas of Thailand: the first officially recognized attack by a communist unit on Thai government troops occurred in 1965 (Bowie 1997: 63). Indeed, it is apparent that the motivations of the projects were at least partially independent of either communist activity in Thailand or even the developmental needs of the population. Consequently, despite the broadly developmental orientation of the US–Thai brand of anti-communism, force remained the arbiter of social conflict in the last instance, and the development approach was, throughout, top-down. As one US economic adviser had put it in a 1956 memo to the Ministry of Finance:

> Economic development needs are not widely felt throughout the country; development schemes do not well up out of long-felt and long-suppressed hopes of regional groups; projects do not originate in the grass roots of the provinces; the provinces are not importuning the central Government to find funds to carry out programs locally conceived. Instead, everything is done by a benevolent bureaucracy in Bangkok. Ministries think up development schemes. Departments of Government decide what is good for the people and then try to get it done (cited in Riggs 1966: 330).

Ten years later little had changed. To be sure, peasants in Thailand had needs that might be met by various development projects. But it was not these needs that drove the development bureaucracy. Rather, the development bureaucracy acted—some members consciously, some less so—as a militarized and class-based promoter of the geographic expansion of internationalizing capital. As such, the bureaucracy might meet some of rural society's needs, but only insofar as this was consistent with the class project of facilitating the peasantry's capture by the forces of capitalism.

2.5 Agrarian change in Thailand under US hegemony

The dynamics of military-led developmentalism can be illustrated by noting some of its outcomes in Northern Thailand, where agrarian change and primitive accumulation were intensified by such developmentalism. Agrarian change in Thailand has been dealt with in great detail elsewhere, and here I will not review the specifics of the process.[49] It will help, however, to briefly summarize the broader context of uneven agrarian change that characterized Thailand as a whole at the point when the events in Northern Thailand described here took place.

Economic geographers and others quite conventionally divide Thailand into five regions: (1) the Bangkok Metropolitan Region (BMR), comprising Bangkok and its five surrounding provinces; (2) the Central region;[50] (3) the Northern region; (4) the Northeastern region; and (5) the Southern region (Donner 1978). (Map 2.1.) While such regional divisions can be simplistic and problematic, these spatial tags can be used here to indicate broad commonalities and differences in the development process across different places in Thailand. As of the 1960s and 1970s, all four of the upcountry regions remained predominantly agricultural in terms of both labour force participation and economic output, but each region exhibited a somewhat different pattern of agrarian transformation. Table 2.1 summarizes data regarding some of these commonalities and differences, based on the *1978 Agricultural Census Report*, the closest census year to the time period of the political events discussed below.

Table 2.1. Agricultural practices, by region, 1978 (% of households)

Agricultural practice	Region			
	Central	North	Northeast	South
Use of 4-wheel tractor	39	36	13	15
Use of 2-wheel tractors	21	8	2	19
Use of engine water-pumps	35	15	6	3
Use of threshers	14	10	1	0
Use of inorganic fertilizers	40	17	19	33
Use of pesticides	49	37	19	12
Irrigation of landholdings	33	38	6	7
Percentage of area irrigated	29	19	3	3

Source: NSO, *Agricultural Census Report* (1978).

[49] For a useful summary of the literature, see Chayan (1993).

[50] In recent years the Central Region has been further subdivided to produce an Eastern Region and a Western Region. In the discussion here, I follow the older, broader definition of the Central Region.

Map 2.1. Thailand and its regions.

Outside the BMR, the seat of power and urban-industrial activity in the Bangkok-based Thai state, the most significant power base for the state has been the Central Region, especially the more densely settled and well-irrigated wet-rice growing areas along the Chao Phraya basin, and the coastal provinces to the east and southeast of the BMR. Commercial transformation of agriculture and substantial agrarian class stratification had already occurred in the Central Region by the 1960s and 1970s, along with some industrialization of agricultural production, development of larger and more concentrated landholdings, and attendant displacement of peasants into non-agrarian occupations, many of these within the BMR. For a variety of reasons, even though class stratification had occurred earlier and to a greater extent in the Central Region, the ongoing agrarian transformation occurring there in the 1960s and 1970s did not lead to the same kinds of intense class-based rebellions as occurred elsewhere in the country (Johnston 1975; Wittayakorn 1983; Douglass 1984).

The South, with a much smaller population than the other three upcountry regions, had also been commercially transformed somewhat earlier than the North or the Northeast, in part because its somewhat different climate and physical geography has allowed more commercial cropping and trade-based livelihood activities, especially rubber production and commercial fishing (Fisher 1964; Fraser 1966). Nonetheless, early commercialization and industrialization was not as intensive as in the Central region. In the South, however, unlike in the Central Region, class and regionally based rebellions became quite powerful by the 1960s and 1970s, even in the context of more restricted stratification and the development of a relatively stable and prosperous group of small traders and local elites. Indeed, the four predominantly ethnically Malay and religiously Islamic provinces of the lower South became important bases for the CPT, and regional autonomy movements continue to be significant in the South up to the present.

Some of the most intensive political struggles against the Thai state in the context of agrarian change took place in the Northeast (*Isaan*). The Northeast is the largest, most populous, most ecologically inhospitable (much land is dry and saline), and poorest region of Thailand. Into the 1960s and 1970s it also showed the most limited development of commercial agriculture. This implies somewhat lesser class stratification in the Northeast than elsewhere in the country, and indeed the modal condition was not for households to be landless but rather to own a small parcel of land, devoted primarily to rice production, that was insufficient to procure subsistence, particularly as economic change accelerated. Shared relative poverty among a large number of Northeastern farmers, and a strong sense of regional identity among the predominantly Lao-speaking residents of *Isaan*, combined with a shared sense of oppression and/or neglect at the hands of Bangkok-based elites to fuel regional political militancy. This militancy was of long standing and had already been exemplified in the 1930s and 1940s by strong support in the Northeast for Pridi's redistributive agenda. As US-Thai counter-insurgency/development policies were aggressively extended into the region during the 1960s and 1970s, this militancy increased, and the Northeast quickly became the major stronghold for the CPT (Morrell and Chai-anan 1981; Girling 1981; Chairat 1988).

In some respects, the agrarian situation in the North lay in between that of the Central Region and that of the Northeast, but the North's situation was also distinctive. Because it is hilly and lightly populated, much of the concentration of landholding and class stratification that did exist in the North was focused along the rich bottomlands of the Ping River (*mae nam Ping*) in Chiang Mai and Lamphun provinces (Table 2.2), devoted primarily to wet-rice production and commercial tree crops such as *lumyai*. Many of these land holdings had long been controlled by well-established elites, some the descendants of Northern Thai royalty from the period when the Chiang Mai-based Lanna Kingdom was an independent tributary of Bangkok (Anan 1989; Turton 1989*a*). While a significant aspect of the political rebellion in the North during the 1960s and 1970s was to be the militancy of ethnically non-Thai 'hilltribe' groups from upland agricultural areas, who had long suffered marginalization at the hands of the Thai state, one of the most powerful political movements of the period developed where the greatest amount of agrarian transformation and class stratification was occurring, within the Chiang Mai Valley. At the centre of this movement was the Peasant Federation of Thailand (PFT), an organization developed by some of the most marginalized peasants in Northern and Northeastern Thailand, with a particularly strong following in the Chiang Mai area.

In the remainder of this chapter, I outline the story of US-Thai collaboration in fomenting capitalist agrarian transformation and in militarily repressing the PFT. This story illustrates how the internationalization of the state has

Table 2.2. Agricultural practices, Chiang Mai and Lamphun, 1978 (% of households)

Agricultural practice	Province	
	Chiang Mai	Lamphun
Use of 4-wheel tractors	12	24
Use of 2-wheel tractors	12	9
Use of engine water-pumps	9	29
Use of threshers	0	0
Use of inorganic fertilizers	21	25
Use of pesticides	45	36
Irrigation of landholdings	76	52
Percentage of area irrigated	69	39

Source: NSO, *Agricultural Census Report* (1978).

functioned in Thailand at a sub-national and local scale. It also establishes a basis for discussion of Northern Thailand's industrial transformation, analysed in ensuing chapters.

More specifically, I focus on two aspects of the Chiang Mai Valley's agrarian transformation: first, the creation of village-level elites through the local penetration of state-led capitalist agriculture; second, the militarization of rural society in response to the social upheavals caused by agrarian change. These phenomena illustrate particularly well the ways in which alliance formation took place within the state around counter-insurgency based development projects. Development, as formulated under the anti-communist rubric of US planners, brought together the interests of the US national security state, the Bangkok-based Thai elites, Chiang Mai-based elites, and a small but burgeoning group of village-level elites in Thai rural society. It also fundamentally pitted these interests against those of the large agrarian population marginalized by capitalist development of agriculture, whether this population was communist in orientation or not. The mix was explosive and required the deployment of US-trained and Bangkok-led security forces to the village to ensure stability.

2.5.1 Capitalist transformation of Northern agriculture: creating allies in the villages

The development of Thai agriculture since World War II has been directly affected by a variety of state activities. Perhaps the most frequently discussed and controversial aspect of state intervention in agriculture has been the rice premium (or rice export tax), maintained between 1955 and 1985. While there has been some disagreement about its effects, a case can be made that it helped squeeze surplus out of agriculture and thus facilitated urban-industrial growth (Silcock 1967*b*; Turton 1978: 115; Medhi 1995: 46–7). At the same time, the premium kept the incomes of both rural and urban producers low, thus balancing urban-industrial growth on the backs of poorly paid rural and urban workers (Chairat 1988: 160–1; Bello, Cunningham, and Poh 1998: 135–6; Dixon 1999: 141–3).

While the rice premium may have helped depress rural and urban wages, the Thai state did attempt to boost agricultural incomes through conventional methods of agricultural modernization, including promotion of higher-yielding varieties, chemical inputs, and mechanization. These efforts were not especially effective, in part because the private credit market has strongly favoured urban-industrial projects over those connected to agriculture, especially in rural areas distant from Bangkok (Kumpol and Panayotou 1985; Nipon and Prayong 1989). While the Thai state tried to overcome these problems through the lending programmes of the Bank of Agriculture and Agricultural Cooperatives and related agencies that developed by the 1960s and 1970s, the main effects of such programmes seem to have been to create a small group of agrarian elites while leaving the majority of agrarian producers marginalized (Anan 1984; Turton 1989*b*; Hirsch 1990).

In the North, the state institutions and projects associated with modernization of agriculture arrived somewhat later than in the Central Region of the country (Wittayakorn 1983; Douglass 1984), but after World War II their presence increased rapidly (Anan 1984). The projects launched as part of agricultural modernization—including the development of new rice varieties and the selling on credit of chemical fertilizers—were directly sponsored and encouraged by agencies such as the US Operations Mission and the World Bank, thus reflecting again the degree to which international collaboration between US and Thai agencies structured the development agenda. These agencies promoted adoption of new agricultural technologies and techniques through a variety of means, including the creation of farmers groups and rural credit cooperatives (Anan 1984; Bowie 1997).

Such efforts did not decrease polarization but, rather, increased it. While agrarian differentiation through increased commodification of land and labour was a process already under way by the late nineteenth and early twentieth centuries, the *pace* of agrarian transformation intensified after Sarit's push for capitalist development took hold, with many peasants being either forced into tenancy or out of agriculture entirely. In the North, during the 1960s and 1970s, there was a highly visible tendency towards smaller average holdings and increased tenancy. Moreover, for those who had been pushed completely out of farming and had to sell their labour directly to agrarian capitalists, the average wage for agricultural labour in the North only reached as high as 5–6 baht per day in 1970, and might be as low as 3 baht (at a time when lunch for one person cost 1 baht). By 1975, these wages doubled, though much of the gain was eroded by inflation, and wages as low as 8 baht per day on larger agribusiness farms were still reported in 1976 (Turton 1978: 114).

The other face of this marginalization of much of the peasantry was a significant consolidation of power and wealth among a small but burgeoning rural elite, many of whom benefited considerably from the largesse made available through counter-insurgency oriented development programmes. Prior to this time, there were two groups of local elites who had gained substantially from the rudimentary capitalist transformation underway in the North: large landowners descended from former Northern noble families, and Sino-Thai merchants (Morell and Chai-anan 1981: 210; Plai-Auw 1986). The further evolution of capitalist development spurred by state projects in the 1960s and 1970s led to greater articulation of the social basis of power in the North—without necessarily changing the percentages of those who could be considered the beneficiaries of capitalist transformation. This happened primarily through a deepening of the circuits of capital within the village, including the development of large agribusiness projects such as the contract farming operations of the Charoen Phokphand Group (Ruangrai and Panayotou 1985; Turton 1989*a*; Hirsch 1990). For rural elites, such changes provided avenues to both wealth and new class status (Anan 1984: 292–7).

Andrew Turton, reflecting on his research in the 1970s and 1980s, (1989*a*: 74) notes that:

State supports seem to be having the effect of differentiating agriculture into two sectors: one advanced, in which agrobusiness and large-scale farming (especially of export crops other than rice) using modern technology is favored, and one backward and disadvantaged, which may at best receive the attention of 'poverty eradication' or 'job creation' programs . . . One might say that in between, and in some ways linking the two, are a minority of producers (though not exclusively producers, but those who are also officeholders, traders, etc.) who seem to be clearly net beneficiaries of limited state economic patronage and, directly or indirectly, state political patronage. And it is often by and through this social stratum that state and parastatal powers are exercised towards the majority of the poorer and disadvantaged rural population.

These 'local powers,' as Turton calls them, constitute roughly 5 per cent or less of the households in each village. Such households 'derive advantage from their external connections and alliances and from their roles in "linking" the majority of villagers with state and market structures. . . .' Through their functions in this role, they are able to 'accumulate (or be at the first stage of accumulation of) village surplus through wages, commodity dealing, retail prices, rent and interest' (Turton 1989*a*: 82).

An important aspect of the village elite in this sort of developmental context is that their activities have by no means been limited to (or even primarily defined by) entrepreneurial innovation or risk-taking. Rather, acting as agents of the state on the 'frontiers' of capitalist development, they have engaged in a variety of forms of primitive accumulation. Turton's composite description of a local elite is revealing in this regard (Turton 1984: 32):

He is a provincial assembly elected representative, owns a transport company, and is involved in sales of agricultural inputs, owns some upland land currently being developed, and is engaged in construction work of all kinds . . . He has various district and subdistrict officials 'in his pocket'. He receives some provincial funds for his 'territory', from which he may take a percentage before passing it on and subsequently also profiting from contracts. He is a member of the local branch of Rotary (or Lions, Jaycees, etc.). He is also chairman of the local Village Scouts and a generous provider of funds for weapons for the local volunteer forces . . . He, or more probably his subordinates, may likely call on 'gun hands' to do their dirty work for them. So he is involved in both capitalist activity and primitive accumulation, with much use of extra-economic coercion. If he were to operate on a big enough scale he might even be termed a local 'godfather', a term which like 'mafia' has entered the Thai lexicon. Indeed the Ministry of the Interior itself, which controls a majority of local officials, is sometimes colloquially referred to as the 'ministry of the mafia'.

The extra-legal and coercive character of local elite activities under counter-insurgency oriented development has been an item much discussed in recent attempts to promote political reform in Thailand (Pasuk and Sungsidh 1994: 57–107). The influence of the political bosses (*jao pho*) has been particularly strong in 'wild east' provinces such as Chonburi (to the east of Bangkok), where the pace and intensity of rapid industrialization has bred a new stratum of gangsterish capitalist politicians (Pasuk and Baker 1997: 182–4).[51] It has also been important in newly burgeoning regional hubs such as Khon Kaen, in Northeastern Thailand (Somrudee 1991), as well as in frontier regions such as the West (Hirsch 1990). However, violence has been integral to development and primitive accumulation not only on the land frontier but in long-settled and densely populated regions—as the experience of the Chiang Mai Valley in the 1970s illustrates poignantly.

2.5.2 *Peasant upheaval and state response: the destruction of the Peasant Federation of Thailand*

The violence that erupted in the Chiang Mai Valley during the 1970s needs to be examined in relation not only to the structural changes affecting agriculture but to the general political-military context. By the 1970s, the strength of the militarized capitalist development process Thailand and the US had helped foment began to show signs of becoming the gravedigger for the militarized version of bureaucratic capitalism. The capitalist class was strong enough to begin challenging the military for control over the political process, while the winding down of US military activities with the withdrawal of troops from Vietnam reduced the leverage and prestige of Thai military leaders. In this context, popular pressures for lifting military dictatorship, led primarily by student organizations, coalesced with dissatisfaction in certain elite quarters to produce a 3-year democratic interregnum. On 14 October 1973, demonstrations led by students were countered by military violence, but this in turn precipitated a split in the ruling coalition, forcing Thanom and fellow military leader Phrapat into exile and opening up a political process that had been dominated by the military and the bureaucracy. One result was an unprecedented movement of business people into cabinet positions (Anek 1988; 1992*a*). Another result was an unprecedented expansion of popular activities by peasant, labour, and student groups, all of which aimed at improving the living conditions of the majority that had been bypassed or subordinated in previous development processes. The latter of these two results was eventually found unacceptable by conservative political elites and business leaders, and consequently the exiled dictators were invited by the Thai monarchy to return to the country (Bradley 1978: 19–22). When students and labour groups protested against this return, the ultimate result was another coup d'etát, this time punctuated by the ferocious and bloody assault on student demonstrators at Thammasat University by right-wing paramilitary groups (Anderson 1977; Bowie 1997: 24–8).

[51] The phenomenon of violence as a tool for gaining political access in democratizing Thailand was opened up for discussion by Anderson (1990), and has been further elaborated by Ockey (1992).

The temporary liberalization of the Thai regime between 1973 and 1976, though focused to a great extent on state institutions in Bangkok, also created some temporary openings for political struggle by rural producers, who could raise political demands in ways that were impossible under the dictatorship. In March 1974, for example, farmers staged a large protest in Bangkok, with support from the National Student Centre of Thailand (NSCT), calling for higher paddy prices. The demonstration was led by a Northern farmer from Phitsanulok province named Chai Wangtaku. The newly installed government, led by Sanya Thammasak (a politically conservative former rector of Thammasat University who had been appointed interim Prime Minister by the King), responded by promising to spend 300 million baht in revenues from the rice premium to implement a policy guaranteeing higher paddy prices (Morell and Chai-anan 1981: 214).

In addition, from March to June of 1974, farmers in the Northern provinces of Phitsanulok, Phetchabun, Phichit, and Nakhon Sawan tried to bring charges against land owners who had loaned them money at higher than legal interest rates. These actions also gained the support of the NSCT, and after some negotiations between farmers, state representatives, and NSCT leaders the government officials agreed to set up a committee that would be empowered to reallocate land and investigate farmers' grievances. This committee solicited petitions from farmers who claimed to have been cheated out of land, receiving an unexpected flood of 10,999 petitions from farmers in 6 Northern provinces during June of 1974 alone. The flow increased in July and August, and by the end of September there had been a total of 53,650 petitions filed, completely overwhelming the committee's capacity. Though it was acknowledged that most of the farmers were filing justified grievances, only 1,635 cases could be resolved by the committee (Morell and Chai-anan 1981: 215–16).

Demonstrations by farmers and student groups continued throughout the year, and one that was held in Bangkok at the end of November may have helped to hasten the passage of a new Land Rent Control Act on 18 December 1974 (Morell and Chai-anan 1981: 217–22; Anan 1989: 101). The act guaranteed tenancy contracts for 6 years, required one rent payment annually regardless of the number of crops grown, and fixed the landlord's share at a maximum of one-half of the crop which was left after deduction for production costs—less than one-half if there was a poor harvest or crop failure (Anan 1989: 101). The act was an important victory for farmers, and especially those from the North, where rents were particularly high. However, the state had little capacity (or perhaps desire) to enforce the provisions of the act, and consequently peasants took it upon themselves to organize an association, the PFT, devoted to promoting rent control and monitoring government implementation of the law, as well as to educating and informing other peasants about the law and its implications (Anan 1989: 101). The PFT was established with headquarters in Chiang Mai on 19 November 1974. Its first president was Chai Wangtaku, and its first vice-president was Intha Sribunruang, a well-known village head (*phu yai baan*), farmer, storekeeper, and head of a Chiang Mai farmers' group (Turton 1978: 122; Morell and Chai-anan 1981: 222).

The PFT established branches in all of the Northern and upper Central provinces and quickly increased its activities in Chiang Mai, Chiang Rai, Lampang, and Lamphun. The high rates of tenancy in these provinces was certainly a major contributing factor to the PFT's growth and it has been estimated that about 60 per cent of all PFT members were poor farmers (many of these tenants), about 20 per cent were landless labourers, and 20 per cent were middling peasants (Morell and Chai-anan 1981: 222–3; Turton, 1987: 38). Yet the PFT also developed a trans-regional and trans-class character. It had substantial backing from student organizations, some journalists and a few politicians, and it reached out to non-peasant groups. It was able to establish branches in 41 of the country's 71 provinces and was strong in the Northeast as well as the North (Morell and Chai-anan 1981: 223; Turton 1987: 38–9; Anan 1989: 101). By mid-1975 one source estimated that the PFT had 100,000 households participating as members in Chiang Mai province alone (Turton 1978: 122; 1987: 37–8; Morell and Chai-anan 1981: 225).

While the PFT had been developed with enforcement of government regulations as one of its main goals, it was distrusted by many state officials. All previous farmers' organizations had been created by the Ministry of Interior or the Ministry of Agriculture and Cooperatives, but the PFT was independent and had been created by the farmers themselves. Officials from the Interior and Agriculture Ministries considered this a threat, and quickly proclaimed the PFT to be an illegal organization on the grounds that it had not formally registered with the government. Intha responded to this charge, in the PFT's magazine, by declaring that the farmers were acting in accordance with the Thai constitution, which guaranteed freedom of association. Furthermore, he claimed, the PFT did not apply for registration 'because we don't agree with the law specifying that people cannot form associations for achieving political ends. This law ignores us and aims to segregate us from participating in the politics of our own country' (Morell and Chai-anan 1981: 223–4). The PFT was thus on a collision course with both elite classes and their representatives in the Thai state, many of whom considered the PFT to be a front for the CPT. While PFT members denied allegations of being a CPT front, the truth probably mattered little to the principal antagonists of the PFT, given the class interests that were at stake in the struggle (Morell and Chai-anan 1981: 224–5). Indeed, the basic conflict was evidenced by the fact that landlords and local officials opposed the Land Rent Control Act from the start and thus the law was implemented on well under 10 per cent of the rice land in Chiang Mai and Lamphun (Anan 1989: 101).

To understand the violence that ensued, it is necessary to briefly revisit the development of the repressive state apparatus that had occurred under US anti-communist tutelage, and particularly the development of paramilitary units such as the BPP and PARU. These paramilitary units functioned under the control of the CSOC/ISOC and the Thai police, and the former of these parent organizations—which was intimately connected to CIA counter-insurgency planning—was to help spawn and legitimize another layer of organizations focused on maintaining security at the grass-roots level. In Bangkok, the best known of these was the Red Gaurs (*krathing daeng*), or red bulls. The Red Gaurs were founded by Special Colonel Sudsai Hasdin, who had formerly been head of ISOC's hill-tribes division (Flood 1977: 5). Among the Red Gaur leaders was his son, Suebsai Hasdin. Other Red Gaur groups were controlled by General Withoon Yasawat, a former leader of CIA-hired Thai mercenary forces in Laos, and General Chatchai Choonhavan, son of Phin, brother-in-law of Phao, and later Prime Minister (Anderson 1977: 28). The organization recruited heavily from the unemployed and from vocational and technical students who felt both threatened by economic downturn in the post-1973 period and embittered at being treated as inferiors by university students, with whom some had cooperated in the 1973 uprisings (Morell and Chai-anan 1981: 241–2). The organization also included many ex-mercenaries hired by various cliques within the ISOC and the police and intelligence sections, and received backing from government servants, soldiers, police, and economically vulnerable middle strata such as small shopkeepers (Anderson 1977: 19–20; Flood 1977: 6). By 1976 it claimed to have over 100,000 members throughout Thailand (Flood 1977: 6).

The Red Gaurs were mainly active in the cities, and their activities will be further mentioned in the next chapter. More important in the extension of paramilitary force to the countryside were two other groups, the Village Scouts (*luk sua chaoban*) and *Nawaphon* (roughly translated as 'new force' or 'ninth power'). The Village Scout movement was established in 1971 by the commanding officer of BPP Region IV (encompassing 15 provinces of Northeastern Thailand), Somkhuan Harikul (Morell and Chai-anan 1981: 242; Bowie 1997: 55–9). The organization received official support not only from the BPP but from the Ministry of Interior, and it was also supported strongly by the monarchy, which encouraged participation in scouting activities as a way of defending nation, religion, and king (Anderson 1977: 20; Bowie 1997: 81–111). At the outset, scout leadership was drawn heavily from the wealthy, the middle-aged, provincial officials, rural notables, and urban nouveaux riches (Anderson 1977: 20; Morell and Chai-anan 1981: 243–4). However, the Scouts were able to become the most extensive mass membership organization on the right by recruiting heavily from the ranks of ordinary villagers, using emotive themes, an elaborate initiation ritual, and various other stratagems to encourage participation (Muecke 1980; Bowie 1997: 183–232). Thus, by 1978, the Village Scouts claimed that over 5 per cent of Thailand's population (or 2.5 million people) had completed the Scout's training programme (Morell and Chai-anan 1981: 243).

The Village Scouts provided an important basis of seemingly broad popular legitimacy for right-wing forces in the countryside, and were an important tool by which local elites maintained their dual positions as clients of higher-level state elites and patrons of lower-level locals. The Scouts also participated in the bloodletting at Thammasat University on 6 October 1976, and were thus important as a paramilitary force (Bowie 1997: 21–33). However, the most important paramilitary force in rural areas may have been *Nawaphon*, which was organized under the ISOC. *Nawaphon* was created in March 1975 by Wattana Kiemvimol, who had been the head of the Thai Students Association at Seton Hall University in the United States (Flood 1977: 6). Wattana claimed to have cultivated contacts with various political leaders and groups while in the United States, including Vice-President Spiro Agnew and the CIA, and upon his return to Thailand he was able to persuade the ISOC to support the creation of an organization targeting residents of provincial and district towns.[52] By late 1975, *Nawaphon* claimed to have over 1 million active members (Morell and Chai-anan 1981: 239).

Like the Village Scouts, *Nawaphon* received the support of the monarchy and had most of its local backing from provincial governors, large landowners, and district and village heads (Flood 1977: 6; Mallet 1978: 84). It also had the support of the popular right-wing monk, Kittiwutto Bhikkhu, who was well known for preaching the merits of killing communists (Mallet 1978: 85; Morell and Chai-anan 1981: 240). Perhaps the most prominent member of *Nawaphon* was the right-wing Chiang Mai judge, Thanin Kraivichien, a favourite of the monarchy who went on to become Prime Minister immediately after the 1976 coup (Flood 1977: 7; Mallet 1978: 84). While *Nawaphon* had its crucial bases of political support within Thailand, it was also alleged by Thai Prime Minister Seni Pramote in 1975 that the organization was funded by 'foreign powers'—which, Thadeus Flood noted, 'could only mean the US CIA' (Flood 1977: 6).

Nawaphon was openly dedicated to polarizing the political situation in Thailand to encourage a coup, and it appears to have engaged in a variety of violent activities designed to further this cause, including attacks on student and progressive journalists and presses (Flood 1977: 6–7). Since much of the violence that occurred during the 1973 to 1976 period was unofficially sanctioned by state officials, there have rarely been revelations as to who was responsible, and thus it is difficult to establish precisely what activities might have been carried out by *Nawaphon* as opposed to other groups. Nonetheless, it has been widely believed that *Nawaphon* was responsible for at least some of the attacks on the PFT that began shortly after the PFT was formed.[53] The best known of these attacks was the assassination of Intha Sribunruang in July of 1975, which occurred one month before the assassination of another Chiang

[52] The former rector of Thammasat University and head of the Bank of Thailand, Dr Puey Ungphakorn, stated that Wattana was asked by the head of ISOC, Gen. Saiyud Kerdphol (who had strong connections with the CIA), to come back from the United States to teach psychological warfare techniques at ISOC (Puey 1977: 11).

[53] *Nawaphon* had become particularly active trying to recruit new members in Chiang Mai and Lamphun at the time assassinations began there in 1975. Other groups suspected of involvement in the assassinations included the army's hunter-killer teams and the BPP (Morell and Chai-anan 1981: 225).

Mai PFT district branch leader. Altogether, between March of 1974 and August of 1975, at least 21 peasant leaders, most of them members or leaders of the PFT, were killed by assassins. Most of the victims were from the North, including eight from Chiang Mai province, two from Phitsanulok, and one each from Lamphun and Lampang. No one was ever found responsible for any of the murders, indicating high-level collusion in the killings (Turton 1978: 123; 1987: 40–1; Morell and Chai-anan 1981: 226–7; Bowie 1997: 155–6).[54]

PFT leaders were not the only targets during this campaign. On 28 February 1976, an assassin killed politician and academic Dr Boonsanong Punyodyana, the Secretary-General of the Socialist Party of Thailand and a supporter of peasant political struggles. Dr Boonsanong, who was from the North, had run for a seat in parliament representing Chiang Mai province zone 1 in 1975. His campaign came under violent attack from mobs of men believed to work for the Red Gaurs or *Nawaphon*. He lost the election, but had significant popular support and continued to be active in the Socialist Party leading up to the 1976 elections. Boonsanong was killed by two gunmen while returning alone to his home in Bangkok after a party, in what appeared to be a professional hit (Trocki 1977: 48–50; Turton 1978: 131; 1987: 42; Bowie 1997: 153–4).

The campaign of terror against poor peasants and their supporters quickly undermined attempts at constructing a viable, legal political opposition. The PFT lost much of its support by late 1976, and the October coup in that year completely closed down avenues for political opposition outside of the CPT (Morell and Chai-anan 1981: 228–30; Anan 1989: 102). Katherine Bowie notes that when she visited Northern villages after the assassinations and asked about the PFT no villagers admitted any knowledge of its existence, the subject having become taboo (Bowie 1997: 156).

In this context, a number of farmers fled to 'the jungle' to fight with the communist insurgents. But while some took up the CPT's call to arms, far greater numbers simply sank back into acquiescence. The CPT in the North was less powerful than in the Northeast, and had much of its support among hill-tribe groups (Race 1974; Saiyud 1985). In any event, the CPT's efforts ultimately failed (Yuangrat 1982; Gawin 1990), leaving the agrarian status quo formed by counter-insurgency oriented modernization very much in place, with important consequences for the nature of the rural industrial development projects that were to come later. In summary, then, the Thai state had succeeded in capturing much of the Thai peasantry and had successfully dismantled opposition to capitalist development in the countryside—including opposition to forms of capitalist development driven largely by the interests of the most wealthy and powerful.

[54] As Turton (1978: 123) notes, the killings 'bore the marks of a Phoenix-type programme adopted from the American experience in Vietnam'.

2.5.3 The geography of class power and marginalization in Chiang Mai: a summary

Counter insurgency in Thailand has often been portrayed as a battle to fend off the growth of communism among the populations considered by Thai and US elites to be the most susceptible to its appeals—those in remote areas and most peripheral to Bangkok-based capitalist development (Saiyud 1985). On this conception, the fundamental issue is that certain groups of people (e.g. the inhabitants of *Isaan* or Northern hill tribes) are marginal because they are 'left out' of development and need to have the benefits of capitalist economic growth extended to them before they side with the opposition.

The explosion of peasant activism and capitalist state repression in the Chiang Mai Valley, however, suggests the importance of a different kind of marginalization: marginalization *within* the process of capitalist transformation. The geographical expansion and intensification of capitalist agriculture, abetted by the internationalized state's specific agrarian policies, produced a marginalized peasantry in the most densely populated region of the North, in the immediate surroundings of the country's major secondary urban centre. This peasantry was marginalized precisely in the process of its 'capture'—that is, the very processes that made it more fully capitalist, integrated into commodity production and commercial economic relations, subjected it to powers over which it had little control.

Simultaneously, the geographical expansion of the capitalist class—through varied forms of state-mediated cooperation between foreign capitalists, Bangkok capitalists, Chiang Mai capitalists, and hybrid rural elites—produced a political alliance that was able to confront this marginalized peasantry effectively with repressive force. Having mastered the space of peasant resistance, this political alliance was able to foment more fully-fledged industrialization within the Chiang Mai Valley by the 1980s and 1990s. Elements of the marginalized peasantry were thus transformed into the members of a marginalized proletariat. In sum, the era of US hegemony saw the geographical expansion to (and within) the North of the classes central to capitalism (capitalists and labourers), a process that was to become more marked with the industrial boom of the 1980s and 1990s.

2.6 Conclusion: the results of the capture of the peasantry

The era of US hegemony saw intensification of the internationalization of capital and the capitalist state. It saw a broadening and deepening of the capitalist accumulation process in Thailand, with the capitalist class in core areas such as Bangkok becoming more developed, and capitalist social relations beginning to take root in peripheral areas where previously their development had been only rudimentary. To be sure, the rural elites who arose from the penetration of the countryside by the Bangkok-based Thai state cannot be described as fully capitalist. As Turton notes, the dominant village strata created by state-led capitalist development 'is now increasingly able to maintain and reproduce itself as part of a small capitalist class, though such a position does not fully or unambiguously define their social location and identity' (Turton 1989*a*: 82).

Yet what is most important here is not the complex, hybrid social forms that evolved at the local level—a phenomenon described in an earlier literature under headings such as the 'articulation of modes of production' (Bruneau 1982)—but rather the fact that the evolution of these hybrid forms was itself part of the gradual expansion of capitalist social relations, driven largely by forces outside the locale, though mediated and refracted by the social forces within it. Furthermore, as Turton notes, 'Through their hybrid class (and non-class) composition these local powers are articulated with the hybrid and heterogeneous power bloc of the state itself' (Turton 1989*a*: 88). Thus, the evolution of local forms of capitalist and quasi-capitalist power can be seen as a function of both the internationalization of capital and the internationalization of the state—in this case through the ongoing mediation of a Bangkok-based sub-imperial project wedded to US counter-insurgency.

Few non-elite Thais were beneficiaries of such internationalist, counter-insurgency oriented development. The destruction of the PFT by the Thai state meant that not only was the social position of peasants more precarious by the end of the 1970s but their political position was weak, with few effective institutional mechanisms of redress. Moreover, as Walden Bello, Shea Cunningham, and Li Kheng Poh note (1998: 151), the destruction of the PFT led not only to short-term disempowerment for people in rural areas but to longer-term destruction of any possible alternative development strategy based on agriculture and more egalitarian distribution of agrarian wealth. Thus, the precariousness which US-backed counter-insurgency/development produced in rural areas was part and parcel of the creation of a development strategy based around urban-industrial growth and a marginalized peasantry that would serve as the source of a politically weak, low-wage urban proletariat. This paved the way for the specific patterns of industrial development that occurred during and after the US period.

Lest this marginalization of the Thai peasantry be too easily taken for granted, we can conclude by noting the important differences that mark development in Thailand and development in Northeast Asia in this regard. In both South Korea and Taiwan, a variety of conjunctural factors led to US and Taiwanese state support for land reform in the 1950s (Cumings 1984; Gold 1986; Jenkins 1991). An important consequence of the substantial reform in Taiwan has been the development of greater production and consumption capacities within rural areas, including the development of relatively robust—and polluting—rural industries (Bello and Rosenfeld 1990). Such relative socio-spatial decentralization of economic power has led to Taiwan's comparatively low levels of economic inequality, as well as to its relatively high spatial decentralization of industrial activity. While the land reform in South Korea was not as thoroughgoing as on Taiwan, and while the geographical distributional effects were therefore less marked, nonetheless in South Korea also agrarian reforms led to a more egalitarian income distribution and a stronger basis for rural industry than in many other NICs (Cumings 1984; Evans 1987; Bello and Rosenfeld 1990).

In Thailand, by contrast, the ability of the US-Thai anti-communist alliance to suppress rural movements for agrarian reform weakened not only the peasantry, but, concomitantly, the prospects for rural industry, thus strengthening the Bangkok-centrism of the economy. In this context, disempowered peasants had increasingly limited options besides migration to Bangkok and entry into the urban-industrial labour force. Destruction of the PFT and successful opposition to agrarian reform by Thai elites and their international allies are thus integral moments in the pattern of Thailand's Bangkok-centred urban-industrial transformation, indicating the different kinds of development choices made by elites in South Korea and Taiwan, on the one hand, and Thailand, on the other. Indeed, the developmental choices made in the Northeast Asian NICs regarding agrarian change have been an important and too often underestimated feature of the creation of Northeast Asian developmental states. The more extreme marginalization of the Thai peasantry, then, was not inevitable or part of a non-negotiable path to modernization but was rather a geographically and historically conjunctural accomplishment of the alliance that helped foment internationalized—and Bangkok-centric—industrial development in Thailand.

3

Internationalization of the State under US Hegemony and Japanese Quasi-Hegemony: Promoting Industrialization and Disciplining Labour, 1945–2000

3.1 Introduction

The fashion in which the Thai peasantry was captured has heavily conditioned the development of the industrial labour process and labour markets. Thai workers did not simply appear at the factory gates when and where they were needed and in possession of the requisite skills. Rather, new streams of marginalized peasants began to join older streams of immigrant Sino-Thai workers as the capitalist transformation of agriculture proceeded, and the ways in which these new streams entered the industrial labour force depended in part upon the ways they were removed from agriculture. Beyond this, the state did not merely passively witness the absorption of former peasants into the industrial labour force but actively abetted the process through a variety of measures, ranging from state promotion of industrial development to investment in education and training of workers. The Thai state also actively shaped the labour market through its alternating suppression and promotion of trade unions, a matter addressed in this chapter.[1]

The state functions that are integral to the industrial transformation described here were carried out by internationalized segments of the Thai state, including one—the Department of Labour—that would typically be associated with national corporatism, thus illustrating the depth and complexity of the internationalization process. The internationalization of capital and the state around industrial manufacturing development has been more

[1] I refer to the range of state actions identified here as part of the process of 'disciplining labour'. As Foucault's analysis indicates (1977), discipline means much more than merely applying the whip and includes a variety of procedures leading to internalization of the values and habits central to industrial labour.

complicated than the internationalization of capital and state in the capture of the peasantry both because of this depth and complexity and because of the overlapping roles played by two hegemons. Whereas the capture of the peasantry was the product of collaboration between Thai and US elites, the disciplining of the industrial labour force involves more multifaceted collaboration among Thai, US, and Japanese elites—as well as transnational statist institutions. Furthermore, there has been some historical phasing of the relative influence of the two hegemons, with US influence declining after the mid-1970s and Japanese influence increasing. Finally, whereas the US intervention in Thailand aimed directly at transforming the structures of state power along with the economy, the Japanese state has been more inclined to make use of the existing state apparatus and to transform its functions, where necessary, through sheer economic power. Thus, Japanese hegemony in Thailand since 1980 is less complete than was earlier US hegemony—following other authors I call it 'quasi-hegemony' (Beeson 2001)—and not only US capital and the US state, but also international institutions such as the IMF, the World Bank, and the Asian Development Bank (ADB), play crucial roles that are to some extent independent of Japanese capital and the Japanese state.[2]

The present chapter outlines how this more complex internationalization of the state has evolved and its consequences for the disciplining of industrial labour. I note the broad social implications of industrial capitalist development in Thailand and the role of the state in relation to the wide range of economic and social requirements for capitalist development. In particular, I use a detailed examination of the transformation of industrial labour and the labour relations system as an illustration of the breadth of the relevant state functions that have become internationalized, tracing the story through both the period of US hegemony and Japanese quasi-hegemony.

3.2 Industrialization, industrial policy, and Thailand as a 'social factory' in the period of US hegemony, 1945–1980

Transformation of the labour process in Thailand under US hegemony and Japanese quasi-hegemony should be seen as part of a much broader industrial/social transformation that affects all facets of life. Thus, the state policies that need to be addressed in understanding industrial transformation are not limited to industrial policy, properly speaking, but include a wide range of policies that affect the availability and quality of labour, as well as the terms on which it is available.

[2] The Japanese, it should be noted, play a more prominent role in the ADB than in the IMF or the World Bank, the latter two having been more heavily influenced by the United States (Potter 1996: 6).

In a discussion addressing this issue, Bell deploys the term 'social factory' to describe the development of capitalist society in Thailand under US hegemony. The evolution of the social factory is based on the fact that capital becomes more broadly social with the evolution of capitalist society and is thus based less on individual capitalists than on the broad, social circuits of capital. As Bell puts it (1978: 53):

> Capitalist society becomes a 'social factory,' encompassing the entire process of social reproduction. Both waged and unwaged workers (and this includes students, housewives, peasants producing for export markets in the Third World, all of whom have been traditionally ignored by radicals) shape the process of reproduction. Moreover, the circuits of capital . . . are international in nature . . . Within the social factory, conceived globally, the task of capital is to establish those social, economic and political institutions which will attempt to ensure the continuous reproduction of these relations. Thus internationally capitalism may have as its goal economic development just as importantly as police repression. Indeed, the relations of capitalist society are guaranteed in the long-run not by physical coercion, but rather by the 'dull compulsion of economic relations', although violence is always used to create a working class.

This suggests that the sort of aggressive militarism evidenced in Thailand during the period of US hegemony is not itself a permanent fixture of capitalism but rather is most (if not exclusively) appropriate to a particular moment of capitalist development—the moment of primitive accumulation (Marx 1967: 713–74). As capitalism has developed in Thailand into the period of Japanese quasi-hegemony, the 'dull compulsion of economic relations' has become increasingly important to capitalist reproduction, while large-scale exercise of violence has become less important.

Yet the social institutions necessary for reproduction of capitalist social relations—even under this dull compulsion—do not automatically issue forth and must be created in processes of class struggle and collaboration. Bell has provided accounts of precisely how this sort of class struggle and collaboration resulted in a broad set of internationalized social institutions within Thailand—all remaining important up to the present—that play integral roles in the reproduction process. Some of these institutions, such as those connected with the agricultural development bureaucracy and counter-insurgency, have already been mentioned. Here I note some of the other institutional developments under US tutelage that have proved important.

First, beginning in the 1950s, the US Operations Mission (USOM) and the US Agency for International Development (USAID) helped train a large number of Thai students in various fields, ranging from the applied sciences to public administration. Between 1950 and 1974, for example, the Fulbright Educational Foundation trained some 8,000 students, most of these in US universities, providing a quarter of the employees in the top four Civil Service classes (Stifel 1976; Bell 1978: 62; 1991: 103).[3] Altogether, up through 1987,

[3] This is in addition to some 18,000 Thai military officers trained in the United States between 1950 and 1980 (Muscat 1990: 20).

USAID funded 11,000 students in the United States and trained 164 out of 411 top Thai state officials (Bell 1991: 103; Muscat 1990: 49–60; Rock 1995: 753). These trainees helped fill out the Thai bureaucracy with personnel who were not merely technically competent but were competent for institutional processes designed to foster capitalist development.

Illustrative here is the case of the economics profession. Economics offerings were expanded at Thammasat University after the abolition of the absolute monarchy in 1932, and a number of faculty were trained in France, where they were influenced by the German Historical School. After Phibun's 1947 coup, however, economists who followed Pridi's political agenda were banished, along with him. During the late 1960s, the Rockefeller Foundation stepped in to shore up liberal economic thought at Thammasat by sponsoring US economists there and by transforming the curriculum, a project which was pushed by the then Dean of the Faculty of Economics, Dr Puey Ungphakorn. Such US intervention in economics departments was not limited to Thammasat but extended throughout the leading Thai universities. One dean of an economics department responded to an interviewer's question regarding how the curriculum at the school was developed, saying, 'we simply look at the U.S. catalogs and make a shopping list' (Bell 1991: 102–4).

The effects of such training in neoclassical and Keynesian economics have not been—and were not intended to be—merely academic. Dr Puey himself was the director of the Bank of Thailand from 1959 until 1971, and many of the Rockefeller-sponsored Thai students who came to the United States during the 1970s returned to Thailand to play leading roles as technocrats and teachers (Bell 1991: 104). Their influence today is felt in everything from interpretations of economic crisis popular in the mainstream media to the actual economic policies of leading state institutions.[4] In this sense, then, those social scientists and state officials who are the product of the internationalized social factory have played a crucial role in reproducing capitalist social relations in Thailand.

Here, it is worth briefly elaborating on the institutional transformations that were triggered by Sarit's 1958 coup. Along with driving home to Sarit the general importance of anti-communism and control of the media, during Sarit's stay in the United States US officials made sure that he heard their views on promotion of private enterprise (*FRUS* 1958–60, xv: 990–5, 1005–8, 1027–30, 1035–8).[5] Sarit acted promptly on these recommendations after his 1958 coup, which took place just after the conclusion of the important 1957–8 World Bank mission to Thailand. The major institutions created or transformed after

[4] Economists are not the only social scientists in Thailand who have been heavily influenced by conservative, Western, intellectual traditions. Hewison, citing Neher, notes that many important Thai political scientists have been trained in interpretations of Thai society stemming from US authors who adopted modernization theoretic and/or structural-functional approaches (Hewison 1997: 4–5).

[5] It is interesting, in this regard, that though Sarit is credited by Thak Chaloemtiarana with having a well developed 'philosophy' sanctioning his 'despotic paternalism' (Thak 1979), he was not seen by US officials in 1958 as having a very clear sense of possible trajectories for industrial development in Thailand. (See e.g. *FRUS* 1958–60, xv: 994).

the World Bank's visit and Sarit's coup included the National Economic Development Board (NEDB, later the National Economic and Social Development Board (NESDB)), the Board of Investment (BoI), the Industrial Finance Corporation of Thailand (IFCT), and the Bureau of the Budget (Silcock 1967*a*: 267–88; Thak 1979: 228–34; Grit 1982: 114–24).

These institutions were very heavily influenced by US and international capitalist conceptions of the appropriate role for state planning in economic development, and were implemented by Thai officials steeped in the sort of training just discussed. The NESDB's plans were consistently drawn up with the assistance of World Bank officials (Grit 1982: 143; Tienchai 1993: 116, 171–2, 351–2). Indeed, the first plan was essentially a repackaging of the recommendations of the 1957–8 World Bank mission (Thak 1979: 228). Consistent with US views of what constituted appropriate strategies for economic development, the plan focused on limiting the expansion of state enterprises and encouraging private investment, particularly by foreign investors, who had the capital necessary to generate a development process based on increased manufacturing output (World Bank 1959: 90–104; Grit 1982: 114–19; Tienchai 1993: 129).

Moreover, the US sent a number of missions to Thailand during the time the NEDB and other central development institutions were being formed or reformed, making known the views of US leaders on the details of how these institutions should be organized and run in order to foster more private enterprise-based industrial growth and increased foreign investment. For example, the US State Department sent an International Cooperation Administration (ICA) mission during 1961 (the 'Bowen Mission,' after mission leader Howard R. Bowen) which recommended specific changes in the structure and approaches of the NEDB, the BoI, the IFCT, and the Bureau of the Budget, including the employment of more US and World Bank advisers to work side by side with Thai planners (ICA 1961: I–4; III–1–3, 6–8; IX–5). The Bowen Mission's report minced no words about the stringency of its requirements, suggesting, for example, that if the Thai government failed to reorganize the BoI and IFCT in accordance with its recommendations, 'ICA-financed projects in the industrial field, specifically those related to the Board of Investment and the IFC/T, be closed out when current contractual arrangements expire' (ICA 1961: IX–3). The seriousness with which the highest levels of US leadership took these recommendations is attested to by the fact that President Kennedy instructed the US Ambassador to Thailand to let Sarit know that 'While [the report] is not a formal expression of United States policy, nevertheless I feel it outlines a sound basis for our future aid activities . . . and will, I hope, become a milestone in our relations.'[6]

[6] Outgoing State Department Telegram number 1006, to US Embassy in Bangkok, containing approved text of letter from President John F. Kennedy to Sarit, 11 Jan. 1962, John F. Kennedy Presidential Library, National Security Files, 1961–3. The US ambassador was to give Sarit a sanitized version of the Bowen Mission report and read the letter to Sarit verbatim.

In addition to the private enterprise orientation this sort of interaction with US officials created in planning offices of the NESDB, the Investment Promotion Laws drawn up by the Thai state and administered by the BoI since the early 1960s have consistently reflected the desires of US and Thai officials to attract foreign capital through various incentives (Grit 1982: 67, 117–18, 123–4, 129; Silcock 1967*a*: 268). The World Bank mission suggested alterations to the 1953 Act on Promotion of Industries, including increased powers for the 'Committee for Industrial Promotion', which was implemented with the ascension to full legal status of the BoI in 1960 (World Bank 1959: 98–100; ICA,1961: IX–5).[7] Following the recommendations of both the World Bank and Bowen Missions, the BoI included on its board of directors the Prime Minister and heads of various powerful ministries.

The IFCT was also set up in response to the World Bank mission's proposals, with the goal of helping fund the development of large-scale local enterprises, though it was also allowed to lend to foreign firms (World Bank 1959: 103–4). At the time the IFCT was set up, the World Bank's International Finance Corporation held 13 per cent of the IFCT's shares, while both foreign and domestic banks doing business in Thailand also held shares. The procedures of the organization, along with its directors, were largely determined by these outside shareholders, ensuring that its loans would be made largely on financial criteria, in spite of the fact that the IFCT was developed partially with public funding and as a quasi-public agency (Silcock 1967*a*: 272–3), which implies that it could be legitimately used as a tool for the attainment of broader social goals.

Finally, along with the NESDB, the BoI, and the IFCT, collaboration between US and Thai elites during the late 1950s and 1960s also led to the formation of the Bureau of the Budget, within the Ministry of Finance, with authority to scrutinize departmental expenditures. It was placed under the direction of the Bank of Thailand, headed during the 1960s by Puey (World Bank 1959: 202–3; ICA 1961: III–3–5; Girling 1981: 81–2). The Budget Bureau has played an important role in enforcing the orientation towards private enterprise-led industrial growth and openness to foreign investment, monitoring and restricting ministerial expenditures not in line with this orientation.

3.3 Labour institutions and the social factory

If the creation of state institutions promoting private investment seems the manifestation par excellence of a capitalist social factory, the development of labour institutions as a manifestation of this same phenomenon cannot be ignored. Indeed, the development of labour relations systems can be seen as

[7] The World Bank mission actually drew up a new investment promotion act as part of its recommendations to the Thai government (World Bank 1959: 98).

integral to the reproduction of specific types of capitalist accumulation projects (Boyer 1994). To comprehend the post-war development of Thai labour institutions adequately, it is necessary to revisit briefly the evolution of Thai labour struggles and the development of the industrial labour process since World War II. After doing this, I proceed to a description and analysis of the development of labour institutions proper, noting how in the internationalization of the Thai state specific sets of labour institutions and labour relations systems favoured by various US and Thai officials came to affect the political and economic position of Thai workers.

3.3.1 Labour militancy after World War II

In spite of the small size and predominantly Chinese immigrant character of the pre-World War II Thai proletariat, working-class struggles in this period were significant and have received some deserved attention in recent years (Skinner 1957; Mabry 1979; Kanchada 1988; Hewsion and Brown 1994). Much of this labour activity was carried out by ethnically based organizations of Sino-Thai workers, but there were also broader attempts to organize and politicize ethnically Thai workers from the 1920s on—including involvement of labour groups in supporting the civilian faction of the People's Party during political struggles of the 1930s.

Labour struggles became more intense after World War II, especially during 1944–7, when Pridi's liberal faction held power. In May 1946, at the instigation of the liberal government, a General Trade Union Association representing both Thai and Chinese workers was formed. By 1947, this federation had been renamed the Central Labour Union (CLU). The CLU was normally headed by Thai political figures but included both ethnically Thai and Chinese unions (Skinner 1957: 286–7; Mabry 1979: 43).

In spite of the urgency of the post-war conditions that stimulated its growth, the CLU was not merely a bread-and-butter labour confederation, and by 1947 it was actively engaged in political activities ranging from organizing 70,000 participants for the May 1947 Labour Day rally to working with left-wing politicians to strengthen labour legislation. Though the CLU was ideologically somewhat heterogeneous, there were a number of Communist Party (CPT) members working within it (the CPT had been legalized in 1942 with the abolition of the Anti-Communist Act, in exchange for the Soviet Union's acceptance of Thailand into the United Nations), and by 1949 the CLU had affiliated with the communist-dominated World Federation of Trade Unions (WFTU) (Mabry 1979: 43; Hewison and Brown 1994: 494–6). In addition, though over half of the CLU's membership originally comprised ethnic Thais, the ethnic Chinese proportion increased over time, and some Chinese unions within the CLU actively supported political struggles in China (Skinner 1957: 287).

The changing global and local political environment of the late 1940s gradually closed down the political opening in which the CLU flourished. Phibun buttressed his anti-communist credentials by gradually cracking down on left-leaning labour organizations. At first, there was no attempt to repress organizations such as the CLU; rather, Phibun attempted to create a separate political base for himself and to widen ethnic splits in the labour force by creating (in 1948) a union exclusively for ethnic Thais, the Thai Labour Union (TLU)—later reorganized and renamed the Thai National Trade Union Confederation (TNTUC). The TLU affiliated with the pro-capitalist International Confederation of Free Trade Unions (ICFTU) in 1950, thus establishing a counterweight to the CLU and its WFTU affiliation (Mabry 1979: 44; Hewison and Brown 1994: 496). By 1952, however, with the intensification of the Cold War and increased US support for the Thai right wing, the CLU was banned and its leaders arrested under an Anti-Communist Act, sponsored by Police General Phao and modelled on US laws regarding un-American activities (Mabry 1979: 44; Hewison and Brown 1994: 496; Thongchai 1994: 6; Fineman 1997: 164–5).

The banning of the CLU did not entirely close down labour militancy, however, and both Phibun and Phao used labour throughout the 1950s as part of a contest for political position. Phao, for example, supported the creation in 1954 of a union to represent those Chinese workers no longer organized since the banning of the CLU. In 1956, Phibun, as part of his broader 'Hyde Park' campaign of political liberalization, supported the creation of a Labour Party affiliated with the TNTUC. During this same period, the Phibun regime also began to establish more comprehensive labour legislation to acknowledge unions and their right to strike, as well as to set guidelines for dispute resolution (Mabry 1979: 44–7; Hewison and Brown 1994: 496–7). Phibun served his own purposes in this move, since he hoped to generate more popular support for his next electoral bid (Phiraphol 1978: 148–50). He also served the purposes of more liberal US elites, who had begun training Thai labour officials in 1955 and who helped draw up the International Labor Organization (ILO) Conventions which Phibun was then being pushed to endorse (Phiraphol 1978: 151–2; Nikom 1982: 118–21; Chokchai 1994: 126). Liberalization, however, allowed leftists within the labour movement more room for manœuvre, and by 1957 they had asserted control within the TNTUC, leading Phibun's regime to form yet another alternative federation, the Thai National Federation of Trade Unions (Mabry 1979: 47; Vichote 1991: 119).

All of these moves were overtaken by Sarit's 1957 and 1958 coups. Sarit established marshal law, banned trade unions, and abolished the 1956 Labour Act (Mabry 1979: 47–8; Wehmhörner 1983: 483). Unions were banned on the grounds that they were 'obstacles to development' and 'gateways for communism to enter Thailand'. Meanwhile, ten labour leaders who had visited China in the 1950s were imprisoned (Wehmhörner 1983: 483). Altogether, 24 labour leaders were arrested and imprisoned throughout Sarit's five years in power, 15 from the TNTUC, while one was executed (Mabry 1979: 47–8; Sungsidh 1989: 23–4).

Sarit's coup was followed by a period referred to as the 'dark ages' for labour, in which all union activity remained officially banned until the early 1970s (Mabry 1979: 49). The absence of official union recognition did not entirely prevent labour activity, however. Between 1958 and 1965 there were 38 strikes, and in the first half of 1965 alone there were 11 strikes involving 3,400 workers. Such activities led to reconsideration of the adequacy of the existing dispute resolution mechanisms and resulted in new labour legislation during 1965 that did not allow unions but refined these mechanisms (Phiraphol 1978: 172; Morell and Chai-anan 1981: 186). This set in train a longer process of transformation—aided by popular agitation and strikes—which finally culminated in 1972 with executive decree no. 103 by the Thanom government, effectively lifting the ban on trade unions (Mabry 1979: 62–3; Morell and Chai-anan 1981: 186–7; Hewison and Brown 1994: 500). The executive decree allowed formation of 'employee associations', but still resisted full union rights such as collective bargaining, political activities, or organization of national federations (Mabry 1979: 54; Morell and Chai-anan 1981: 187; Wehmhörner 1983: 483; Sungsidh 1989: 11–12).

The ousting of the military dictatorship in 1973 created a political atmosphere more conducive to labour activity. In the period from 1973 to 1976, there was an unprecedented outpouring of labour militancy, with the number of strikes reaching 241 in 1975 (Girling 1981: 201–2; Wehmhörner 1983: 483). The Thai government attempted to contain this militancy by further institutionalizing labour relations under the Labour Act of 1975, based loosely on the US Labor Relations Act of 1935, which allowed collective bargaining and the formation of federations (Nikom and Vause 1994: 20). In spite of these provisions, labour militancy did not subside until conservative elites reasserted control through the military coup of 1976 (Wehmhörner 1983: 484; Hewison and Brown 1994: 500–1). In the aftermath of the coup, labour leaders were arrested and strikes banned, and at least 50 peasant, student, and labour leaders were assassinated by right-wing vigilante groups, such as the Red Gaurs (Flood 1977: 5–6; Anderson 1977: 19–20; Morell and Chai-anan 1981: 241; Sungsidh 1989: 203). The right-wing repression gradually lifted by 1980, but the routing or co-opting of the left within the labour movement had largely been achieved by that time (Sungsidh 1989: 207–50).

3.3.2 The role of the US in the development of Thai labour relations

An under-investigated feature of these post-World War II events is the role of US agencies in shaping Thai labour relations. The repressiveness of the Sarit regime was clearly encouraged and supported by the United States, but there was no uniformly anti-labour union stance on the part of the US imperial state. Repression was only one of several approaches to containing labour militancy, and was supplemented in various times and places by attempts to institutionalize labour conflicts and build conservative, anti-communist trade unions (Sungsidh 1989: 205–11).

Liberal internationalists had been predominant within the US foreign policy establishment in the immediate post-war period, until Truman's ignition of Cold War rhetoric in 1947 began to push the entire US political spectrum further to the right. The originally more liberal orientation is reflected in the comparatively subdued, though obviously concerned, comments by US Ambassador Stanton regarding the role of communists in the 1946–7 strikes. The Ambassador noted matter-of-factly in March of 1947 that the rise of CPT and labour activity might prompt Phibun to attempt to return to power.[8] In a June 1947 fortnightly summary of events discussing the May Day labour rally attended by some 70,000 workers, he also noted without much comment the role of the CPT and its government sponsors. In conformity with the standard US view of Thai communism as a basically Chinese phenomenon, Stanton noted that it was unclear to what extent communists were involved in the labour movement and that those members who were Siamese were deemed 'of a more moderate type, utilizing the appeal of communist ideology more as an ultimate goal than as an immediate objective'.[9] Even though Pridi, whose government had promoted labour unions and activities, was seen as possibly encouraging communism in Thailand, he was portrayed as unlikely to align with the Soviet Union.[10] And in discussing a strike at Shell and Standard Vacuum Oil Companies on 18 September 1947, Stanton acknowledged that promotion of this strike by leftists within the government was in conformity with principles of the ILO, which Thailand had joined.[11]

Nonetheless, even the least anti-communist US officials cannot have been upset with Phibun's decision to promote conservative unions against the CLU after he regained power. A 1949 CIA report on communism in Asia warned that 'Communist movements derive their principal support from urban labour in Japan, India, and Thailand', indicating that Thai labour was to be a focus of concern.[12] USOM thus largely acquiesced in whatever repressive policies Phibun adopted towards labour, but there were also some small beginnings of an effort to build a Labour Department and to put conservative labour unions more firmly on the agenda.

Labour affairs had first been institutionalized in 1933 with the creation of a small Labour Division within the Department of Public Welfare in the Ministry of Interior (Chokchai 1994: 94), and by the 1950s such affairs were still attended by this small and technically inadequate division (Sungsidh 1989:

[8] 26 Mar. 1947 telegram from Stanton to Secretary of State, US National Archives, State Department Files on Thailand, 1945–9, 892.00.

[9] 7 June 1947 fortnightly summary, US National Archives, State Department Files on Thailand, 1945–9, 892.00.

[10] Ibid. 26 Aug. 1947.

[11] Ibid. 9 Oct. 1947.

[12] Central Intelligence Agency, 'Intelligence Memorandum No. 208, Communist Methods in Asia', 26 Aug. 1949, Declassified Documents Reference System, 1976, 3C: 3.

24). As a US Department of Labor study noted, Thai unions in the 1950s 'had no defined legal status or legal protection and were apt to be dealt with arbitrarily by employers or by the police' (US Department of Labor 1959: 21). To begin redressing this problem, USOM began in the 1950s to train a small number of Thai officials in the United States in order to build the capacity for development of a more fully-fledged state labour bureaucracy that could better define and regulate the status of labour organizations. Thus, during the years when Phibun was liberalizing the political arena (1955–7), USOM trained 14 Thai officials in the area of labour affairs, including the Director-General and Deputy Director-General of the Department of Public Welfare, the Chief of the Labour Division, and the Chief of the Labour Survey and Research Section of the Division (ICA 1956: 22–3). To be sure, this training was a very minor part of USOM's overall operations, constituting less than 1.5 per cent of all trainees between 1951 and 1957, and the much greater number of specialists trained in areas such as agriculture, industry, transportation, health, and public administration shows the relative lack of concern the US mission had for labour issues (Table 3.1). Yet USOM was at least devoting some resources to this task, and as a later Director-General of the Department of Labour noted, had USOM's support not been interrupted by Sarit's coup, Thailand might have had a fully-fledged Labour Ministry some 30 years earlier than it eventually did.[13]

At the same time as USOM was training labour officials, the ICA and the US Information Service were providing support to Phibun's attempt to build conservative unions by supplying TNTUC with printed materials, movies, and loudspeakers (ICA 1956: 7). The American Federation of Labor-Congress of Industrial Organizations (AFL-CIO) took initiative also, sending Vice-President William C. Doherty to Bangkok in July 1956 to build contacts with TNTUC and the Ministry of the Interior, and later the AFL-CIO sponsored at least one TNTUC unionist to come to the United States for a summer seminar.[14] Meanwhile, the US-backed ICFTU sent representatives to Thailand and trained Thai labour leaders abroad between 1950 and 1958 (ICA 1956: 8).[15] Even after Sarit's 1958 coup, some officials continued to consider building conservative unions to be a viable alternative to banning them outright. A fundamental problem for the Thai leadership had been that though they effectively outlawed leftist unions in 1952 through the Anti-Communist Act, their

[13] Author interview, Dr Nikom Chandravithun, 4 June 1997.

[14] 'Memorandum of Conversation', 23 July 1956, on meeting between AFL-CIO Vice-President William C. Doherty and Thai trade unionists, and letter from Anthony Clinton Swezey, Acting Officer in Charge of Thai Affairs, to John I. Getz, Second Secretary of Embassy, American Embassy, Bangkok, 29 Aug. 1958, both in US National Archives, RG 59, Bureau of Far Eastern Affairs, Subject Files Relating to Thailand, 1955–9, Lot File 60 D 50, Box 1, Labor Matters.

[15] 'Discussion of Various Labor Matters', 17 June 1958, US National Archives, RG 59, Bureau of Far Eastern Affairs, Subject Files Relating to Thailand, 1955–9, Lot File 60 D 50, Box 1, Labor Matters. The ILO also trained a number of Thai, both in-country and abroad, throughout this period (ICA 1956: 10–11; US Department of Labor 1959: 29).

Table 3.1. Participant training under USOM-supported programmes in Thailand, 1951–1971 (number of persons trained)

Year	Field of training						
	Agriculture	Industry	Transport	Labour	Health	Education	Public Administration
1951	12	2	3	0	61	0	0
1952	28	15	8	0	8	7	18
1953	37	19	14	0	41	26	10
1954	74	15	18	0	79	81	21
1955	32	13	26	3	59	58	25
1956	37	17	35	4	33	68	29
1957	59	17	36	7	38	124	37
1958	59	2	22	0	25	95	55
1959	153	9	16	0	24	168	52
1960	113	8	40	0	25	82	39
1961	152	9	20	0	32	82	32
1962	59	16	6	0	26	127	20
1963	93	11	12	0	29	87	36
1964	116	0	0	0	54	20	26
1965	114	10	3	0	93	7	55
1966	49	24	4	0	129	95	101
1967	152	10	8	4	88	91	119
1968	124	4	12	10	116	114	97
1969	153	21	22	9	99	174	72
1970	138	39	18	7	132	157	96
1971	183	28	15	2	136	52	9
Total	1,937	289	338	46	1,327	1,715	949

Source: USAID (1972).

chosen tool of co-optation, the TNTUC, was gradually pulled to the left itself as it became the only legal avenue for worker struggles against capital (US Department of Labor 1959: 23–4). An ICFTU official who visited the US Embassy in December 1958, characterized the (by then banned) TNTUC as 'leftist-oriented', but still encouraged US officials to support a strategy of encouraging individual industrial unions as a counterbalance, since individual unions could be headed by local working leaders and would be less prone to being taken over by non-working leftist politicians, as federations like the TNTUC allegedly had been. This approach seems to have met with general approval from the US Embassy staff.[16] Whatever might have been, however, was firmly derailed by the realities imposed under Sarit. USOM stopped training labour officials in 1958 and there was no more such training until 1967.

Yet the complete suppression of all labour activity was impossible and, as many officials recognized, there were problems with trying to maintain control of labour through use of the stick alone. Thus, by the mid-1960s, there was once again talk of developing conservative trade unions. A 6 August 1965 telegram from the US Embassy in Bangkok to the Secretary of State commented that the new labour bill passed by the cabinet in that year would seem to allow in the future for 'labor organizations carefully controlled by the government'.[17] Meanwhile, a joint Thai–USOM task force had in 1963 recommended upgrading the Labour Division to a full department within the Ministry of Interior and in 1965 this upgrading occurred (Halm 1973: 1; Chokchai 1994: 94). Subsequently, the USAID followed up hints of liberalization by sending a labour consultant to review the newly elevated department's operations, and this consultant's recommendations resulted in a programme designed to encourage development of legal labour unions and more comprehensive labour legislation (Halm 1973: 2).

The programme, called USOM's 'Labour Department Administration Project', began in 1967 and continued until 1974. It was followed by a 'Labour Training and Management Project' in 1972–5. USAID spent US$435,000 on the 'Labour Department Administration Project' and another US$190,000 on the 'Labour Training and Management Project'—small totals in comparison to USAID's overall spending in Thailand, but adequate to help create a stronger institutional basis for labour regulation (USAID project database). The money was spent on a variety of activities, including training Labour Department personnel in labour administration, a total of 39 such people being trained in the US between 1967 and 1972 (Halm 1973: 3). Trainees included each of the Department's three Directors-General over the period, along with the Deputy Director-General, all of whom were brought to the

[16] 'Memorandum of Conversation, New ICFTU Delegate Discusses Position Here', meeting held at American Embassy in Bangkok, 31 Dec. 1958, US National Archives, RG 59, Bureau of Far Eastern Affairs, Subject Files Relating to Thailand, 1955–9, Lot File 60 D 50, Box 1, Labor Matters.

[17] Telegram no. 221, from American Embassy in Bangkok to Secretary of State, 6 Aug. 1975, Declassified Documents Reference System, 1975, 884 C: 3.

United States for training (Halm 1973: 3–4). In addition, the Labour Training and Management Programme offered 7 courses to 228 persons between 1972 and 1975 (USAID project database).

The USAID project, which was also assisted by the United Nations Development Program and the ILO, encouraged the Labour Department to devote 'a good deal of time and energy . . . to the development of legislation which would modernize the then existing labour laws', according to US participant Glenn E. Halm (Halm 1973: 4–5). As such, the project contributed to Thanom's National Executive Council Announcement no. 103, in March 1972, allowing for the organization of worker 'associations'. Indeed, USAID project reports take credit for the passage of both the legislation that came into effect with the 1972 decree and the 1975 labour legislation (discussed below), citing these as outcomes of the projects (USAID project database).[18]

The reasons for such a move by the United States are not hard to discern. On the one hand, the development of a more complex industrial society requires more complex methods of disciplining labour and ensuring not only its subservience to capital but the quality of its work performance. Labour unions and more comprehensive labour legislation can be useful tools in procuring these outcomes, provided that the unions are adequately controlled by the state and behave 'in a responsible manner', as US labour officials believed was the case in early 1973 (Halm 1973: 9). Furthermore, labour unions can potentially provide controlled venues for expression of worker grievances. As Halm noted in 1973 (29–30):

> It would be surprising if the next decade was characterized by labor peace and tranquility. The past fifteen years in which labor had no effective legal means to rectify abuses, real or imagined, has built up a backlog of complaints. Announcement #103 provides a means for, and tacit approval, of redress for these complaints. It seems probable that in the immediate future workers will energetically seek to improve their condition and to test their new rights. This activity should recognized [*sic*], noisy and troublesome though it may be, as a generally healthy release of pent up pressures through the 'safety valve' of Announcement #103. If this legal release of pressure were not available to workers, more explosive reactions would seem predictable in the foreseeable future. This seems particularly true since the recent rise in the cost of living has added a new dimension to the worker's difficulties.

Popular pressures and workers' grievances were indeed being felt before the enactment of Announcement no. 103, making the recommendations of the US officials credible, and the strategy of buying worker support through a minimum wage—first enacted with the 1972 decree—more palatable to Thai elites (Phiraphol 1978: 173–81; Sungsidh 1989: 110–12, 120–1).

This strategy should also be seen against the backdrop of changing US commitments to Southeast Asia. By 1968, the United States was beginning to

[18] A former USAID official told the author that the US State Department encouraged Thanom to lift the ban on unions. Author interview, 18 July 1997.

reduce the use of ground forces in Vietnam, as part of what became known as the process of 'Vietnamization' (Kolko 1985: 312–55; Young 1991: 210–31, 240). In general, from the late 1960s on, the US government was inclined to leave more of the burden of defence against communism to local governments themselves, and for an industrializing society such as Thailand this suggested the need for more sophisticated and less uni-dimensional approaches to labour control. The Labour Act of 1975, which still provides the basic framework for labour relations, was a means to such sophisticated labour discipline, drawn directly from US models (Nikom and Vause 1994: 12–66, 127–69; Chokchai 1994: 108).

The 1975 Act allows collective bargaining by unions of either the enterprise or industrial type, but it requires that all union members be Thai nationals and that unions be promoted by at least 10 non-probationary workers who are Thai nationals and are over 15 years of age (Nikom and Vause 1994: 30–1, 157). The requirement hinders unions from two directions. On the one hand, it can make unions difficult to organize in the first place, because many factories hire very young workers or non-Thai nationals and many factories also keep workers in the probationary stage of employment for at least six months.[19] Furthermore, the Labour Department, having been given no enforcement power by the act, does little to prevent factories from firing union promoters during or after the time they attempt to gain the support of at least 10 people, thus inhibiting organizing drives (Sungsidh 1989: 207–8).

On the other hand, where organizers are able to overcome these hurdles, the regulations pose further problems, since they allow unions to contain as few as 10 people and allow for creation of more than one union even within the same enterprise. They also allow as few as 15 unions to form a national labour congress. As a result, the labour movement has been highly subject to fragmentation and has become an arena for narrowly self-interested individuals to compete for opportunities to administer patronage and enrich themselves (Chrobot 1996). To cite examples of the effects of this legislation, in 1988 there were only 56 state enterprises but 115 state enterprise unions (Chokchai 1994: 111). The Port Authority had 6 different unions in 1990, while the Bangkok Mass Transit Organization by itself had 22 unions. The textile and garment unions set up two competing Textile and Garment Federations (Sungsidh and Kanchada 1994: 225; 1996: 22). Fragmentation and low union density has been even worse in the private sector; by the mid-1990s, state enterprise unions had an average of 2,769 members, while private unions averaged only 269 (Sungsidh and Kanchada 1996: 22).

Fragmentation and extremely low levels of overall unionization are only part of the problem of union weakness in Thailand, however. Among the other important problems is the fact that stiff private employer opposition to union organizing has contributed indirectly to strong social and geographic strati-

[19] Author interview, Thai labour organizer, 13 and 18 May 1997.

fication of the labour movement, which makes national-level, collective action in the interests of all workers very difficult to generate. Employer resistance to unions is similar to what exists elsewhere in the developing world, and in Thailand companies resistant to unions are frequently able to fire promoters with impunity before the union can be organized (Brown and Frenkel 1993: 90). Since major branches of the state have generally supported attempts to weaken labour, while the Labour Department has had no enforcement power, workers rarely gain state backing in claims of unfair dismissal (Sungsidh 1989: 86, 207–8, 212). Thus, between January and September of 1975, the Department of Labour recorded 8,100 dismissals of workers from 1,931 enterprises. Most of these were private sector union activists and leaders (Sungsidh 1989: 233). According to one study, between 1975 and 1993, a total of 10,000 union promoters and leaders were fired by management (Pasuk, Sungsidh, and Nualnoi 1996: 107).

The difficulties of forming unions in the private sector meant that by the late-1970s the majority of unionized workers were in state enterprises, even though most of the labour force was working in private factories (Sungsidh 1989: 221). One result of this was the development of a highly stratified labour movement. The state enterprise sector was relatively well organized, had some support within the state, and had the sheer size and legal expertise to ensure that at least some of the existing labour laws were actually implemented in state factories. The private sector unions, by contrast, were smaller and weaker, and state enterprise workers did not always extend solidarity to these unions—something that was in any event made difficult by the fact that labour laws prohibited 'political' unions and secondary strikes (Nikom and Vause 1994: 157–64).

Furthermore, state enterprise workers have overwhelmingly been men, while most of the workers in key export industries of the private sector have been women (NSO, *LFS* 1983–2000; Sungsidh and Kanchada 1996: 243–7). Thus, even though Thailand has very high rates of female labour force participation—with the proportion of the formal wage labour force constituted by women usually running just under 50 per cent throughout the post-war period and the percentage of women and men employed in manufacturing being nearly identical by 1990—women have been highly under-represented in unions, labour federations, and labour congresses. For example, in 1993 women constituted just 27 per cent of the membership of 54 surveyed unions, and only 30 per cent of the leadership of those unions. They also constituted just 14 per cent of the leadership of the 18 labour federations, and 19 per cent of the leadership of 7 national labour congresses (Pasuk, Sungsidh, and Nualnoi 1996: 92–4). Moreover, unions were present in only about one per cent of all Thai garment firms—which employed about 800,000 workers in 1992, 85 per cent of these women—and the predominantly male state enterprise union leadership has often done a poor job of addressing many issues of significance to these female workers (Deyo 1995*b*: 27, 29).

Finally, most of the state-enterprise unions were, and still are, located in and around Bangkok. While this has been true of the private-sector unions as well, the existence of larger numbers of private firms throughout the provinces makes attention to provincial labour conditions more important to private-sector unions than to the state-enterprise unions (Mabry 1979: 71; Wehmhörner 1983: 485–6; Sungsidh 1989: 122, 129, 185, 234–5). In sum, then, the state enterprise unions historically became a small, male-dominated, Bangkok-centric 'labour aristocracy'—albeit of a limited sort—which has sometimes indirectly abetted the fragmentation of the broader labour movement, reinforcing vertical alliances between labour and specific employers rather than encouraging horizontal alliances between workers (Sungsidh 1988: 65; Sungsidh and Kanchada 1994: 243–7; 1996: 24–5; Pasuk and Baker 1995: 189–91).

It is important to note that such an outcome was facilitated by the activities of both Thai and US union backers. The Thai state followed a strategy of opposing and undermining the student-backed Labour Coordination Center of Thailand (LCCT), while promoting the somewhat more conservative Federation of Labour Unions of Thailand (FLUT), in which the state enterprise unions played a leading role. The politics of these union federations, which were first allowed to form by the 1975 Labour Act, were not simple, particularly given the complex political context in which they came to birth. FLUT, for example, contained both private and state enterprise unions, of varying degrees of political militancy, and some FLUT members, such Arom Pongpangan, had good relations with the more radical student leaders. In general, however, it is fair to say that FLUT tended to steer clear of the militant political stance taken by student activists and LCCT members, sticking to a more 'bread-and-butter' form of unionism (Morell and Chai-anan 1981: 191–3; Wehmhörner 1983: 486–7; Sungsidh 1989: 142–253). FLUT's stance was not surprising, given that it had in fact developed out of a labour association formed with the backing of the Labour Department in 1972 (Sungsidh 1989: 220). Its leaders were trained by representatives of the German Social Democrat Party's Friederich Ebert Stiftung (FES) and the AFL-CIO's Asian–American Free Labor Institute (AAFLI) (Sungsidh 1989: 226).

The significance of AAFLI's role during this period is a neglected topic and is difficult to investigate, since AAFLI does not make its files available to researchers.[20] What is known about AAFLI's activities during the 1970s is that it first set up an office in Bangkok during 1975, just as the 1975 Labour Act was put into place.[21] From its inception, AAFLI was mired in controversy over

[20] My request to see AAFLI's files on Thailand in July 1997 was turned down. Other researchers have had similar experiences with the AFL-CIO's foreign policy institutes. See Sims (1992: 3–4.)

[21] Author interview, former AAFLI official, 18 July 1997. AAFLI was founded in 1968 and had already been active in places such as Vietnam and the Philippines before coming to Thailand. Its activities in those places has been somewhat better documented. See e.g. Siegel (1975); Spooner (1989); and Sims (1992).

allegations that it worked as a CIA front, but the truth or falsity of these allegations matters less than the fact that the liberal wing of the CIA and the anti-communist leadership of most US labour organizations saw eye-to-eye about the desirability of developing pro-capitalist labour unions.[22] A former AAFLI director stated in an interview that AAFLI had focused fundamentally on training leaders of the state enterprise unions, which is consistent with the Thai state's policy of creating an alternative to the LCCT and the private unions.[23]

To be sure, there are a variety of reasons that could be given for AAFLI to focus on the state enterprise unions, including the fact that they were already fairly well established and supported by the state, and with their more highly educated leadership could potentially serve as the most effective seeding grounds for development of model labour unions (Wehmhörner 1983: 484–5). Yet this has not happened in Thailand. Instead, the relative lack of support for unions in the private sector and the state's repression of radical alternatives have led to the stratification of workers and weak private-sector unions, including many single-firm, enterprise unions with no connections to other labour organizations. The failure of international labour organizations to adequately support organizing drives in the private sector has thus prevented the public sector unions from having any demonstration effect. Furthermore, while US-based labour organizations such as AAFLI can cite the prevalence of large industries in Bangkok as a good reason for focusing organizing efforts on that region, their failure to embrace a broader agenda and bring organizations such as the Peasant Federation of Thailand (PFT) into organizing drives in effect worked to support the strategy of creating a Bangkok-centric 'labour aristocracy'. The student movement and its National Student Centre of Thailand (NSCT), by contrast, worked to bring groups such as the LCCT and the PFT together into a broad national political movement (Napaporn 2002). The state's attacks on these organizations can thus be seen as one side of a coin, the other side of which is the liberal internationalist strategy of labour co-optation, making horizontal affiliation of the sort the NSCT desired impossible and the more vertically oriented and fragmenting affiliations favoured by organizations such as the state enterprise unions the only viable option (Sungsidh 1988: 73–4; 1989: 142–71).

After the 1976 coup, the attempt to institutionalize and co-opt labour struggles briefly took somewhat of a back seat to out-and-out repression of the Left. Even some of the union leaders within FLUT became targets, and Arom was

[22] Bevars Mabry, whose late 1970s account was for many years one of the few, extensive, English-language studies of Thai labour, asserted that he believed AAFLI was not acting as a CIA front in Thailand (Mabry 1979: 74). In contrast, Philip Agee, a former CIA operative in Latin America, has claimed that AAFLI in general was permeated with CIA agents (Sims 1992: 60). A former CIA operative, who worked for the CIA in Thailand during the 1960s, told me during an interview (13 July 1997) that while he was not aware of whether or not AAFLI-Thailand worked with the CIA, such promotion of conservative labour organizations was precisely the sort of activity the CIA favoured.

[23] Author interview, former AAFLI official, 18 July 1997.

thrown in jail for two years, along with 17 student leaders (Morell and Chai-anan 1981: 192). But nonetheless the clock was not entirely turned back to the days of Sarit. The new regime, though it banned strikes, negotiated an agreement with the conservative faction of FLUT, thus keeping the co-optation option open (Morell and Chai-anan 1981: 200; Bello, Cunningham, and Poh 1998: 76). After Kriangsak Chamanond's coup removed Thanin from power in 1977, the state acted to institutionalize labour disputes further by refining the dispute resolution process under the 1978 Labour Reform Bill. The bill established a National Labour Development Advisory Committee, fashioned after the US National Labor Relations Board, a Labour Relations Committee, and a Labour Court (Prizzia 1985: 33–6; Nikom and Vause 1994: 57–9). The Labour Court, which is empowered to hear complaints by either employees or employers regarding violations of labour laws, was modelled on European systems and thus exemplifies another nuance in the internationalization of the state.[24]

These manœuvres did not eliminate strikes, which continued to occur on a wildcat basis, but they did help elaborate a set of state mechanisms for channelling worker discontent in ways that were containable. By 1981, the dual process of suppressing leftist unions while promoting more conservative unions and channelling their activities had evolved to the point that the ban on strikes could finally be lifted (Wehmhörner 1983: 484; Hewison and Brown 1994: 501). What Thailand was left with by the 1980s, then, was a highly constrained, highly fragmented, and partially institutionalized labour movement. Liberals hoped to see this movement evolve into a more full-blown set of labour-management mechanisms that would resemble those of core-country Fordism, while more conservative elites were simply satisfied that for the time being labour militancy seemed contained. By the 1990s, both groups would find their hopes somewhat undermined by events.

3.4 Overview of Thai political economy under Japanese quasi-hegemony, 1980–2000

While US support was critical in helping lay many of the foundations from which the contemporary Thai state and economy developed, it has nonetheless been capital from Japan more than from any other foreign country that has contributed to recent Thai economic growth and industrialization, abetted somewhat since the end of the 1980s by new investments from Taiwan. As much of the literature on Japanese foreign policy observes, there is somewhat of a mismatch between Japan's regional and global economic power and its international political influence (Sadao 1989; Beeson 2001; Hook 2001;

[24] Author interview, former Labour Court justice, 3 Mar. 1997.

Susumu 2001). While Japan has emerged as the predominant economic influence throughout Northeast and Southeast Asia in recent decades (Hatch and Yamamura 1996; Tabb 1995), it maintains a very low political profile, in part because of its relative political and economic subordination to the US government since World War II (Cumings 1999*a*; Johnson 2000) and in part because of the Japanese government's own concern to minimize the potential for nationalist backlash against the Japanese investment in the region—a backlash that in fact had already occurred in Thailand by the mid-1970s. Nonetheless, Japanese foreign policy has long been directly supportive of overseas Japanese corporate ventures, through such mechanisms as Overseas Development Assistance (ODA) to countries that form joint ventures with Japanese firms (Potter 1996; Steven 1996; Hook 2001). Moreover, organizations such as the Japanese External Trade Organization have also played a direct government-to-government role in encouraging openness to Japanese investment.

Japanese economic relations with Thailand have also been facilitated by well-established relations between various Japanese and Thai state officials, some dating back to the World War II era (Doner 1991: 79–80). Though these relations were held somewhat in abeyance during the period of US hegemony, Japan had already re-emerged as the dominant foreign economic force within Thailand by the 1970s (Unger 1998: 117–18), and its importance has continued to grow since then—at least up to the start of the economic crisis that began in 1996–7. The growth of Japanese investment has both spurred and reflected the development of a more export-oriented regime since the 1970s. The ascendance of this regime is based on the relative decline in strength of the military bureaucracy and the state enterprise sector along with the increasing power of both (global) export-oriented Japanese transnational corporations (TNCs) and export-oriented (but sometimes investment-constrained) Thai capitalists.

The economic dominance of Japan and the relative decline of the United States are both reflected in trade and investment figures for the Thai economy. The importance of Japanese imports has been paramount since the 1960s, accounting for 32.2 per cent of Thai imports in the 1960s, 31.4 per cent in the 1970s, 26.9 per cent in the 1980s, and 30.2 per cent from 1990 to 1995. US shares of Thai imports have decreased steadily over this period, from 20.2 per cent in the 1960s to 11.6 per cent during 1990–5 (UN, *International Trade Statistics Yearbook* 1965–97). Figures on FDI in Thailand also illustrate the declining relative significance of US-based investors as a source of capital for Thai manufacturing, and the increasing relative dominance of Japanese investment: during 1965–72, Japanese investors were responsible for 28.8 per cent of all FDI in Thailand, compared to 45.1 per cent for US investors, but by the 1987–95 period, Japanese investment had risen to 31.6 per cent of the total (in spite of large inflows from elsewhere in Asia), while the US share had declined to 13.2 per cent (Bank of Thailand, *Annual Report* 1975–95).

BoI approvals confirm this trend, which has been particularly evident since the mid-1980s. After the 1985 Plaza Accord, which forced a revaluation of the Japanese yen, and the consequent onset within Japan of the 'high yen recession', Japanese capital began to flow into Thailand in record amounts. From 1988 to 1995, the total value of BoI-promoted projects that started with Japanese investment was 208,178 million baht, while the registered Japanese capital of BoI-promoted firms opening during this period amounted to 53,376 million baht. By comparison, BoI-promoted projects with US participation that opened during the period had total investment amounting to 68,598 million baht, while the registered US capital in these projects came to 6,264 million baht (BoI *Investment Statistics*, 1988–95).

During the economic boom, Thailand also began to receive substantial FDI from Singapore and the Northeast Asian NICs, particularly Hong Kong and Taiwan. Thus, in 1994–6 Hong Kong accounted for 14.5 per cent of all FDI in Thailand, Singapore accounted for 10.7 per cent, and Taiwan for 5.7 per cent. Yet during this same period, these countries played a much smaller role in Thailand's trade, with Hong Kong taking 3.6 per cent of Thai exports and providing 1.2 per cent of Thai imports, Singapore taking 9.4 per cent of Thai exports and providing 5.9 per cent of Thai imports, and Taiwan taking 2.4 per cent of Thai exports while providing 4.7 per cent of Thai imports (Bank of Thailand, *Quarterly Bulletin*, December 1997). Thus, in spite of the growth of trade and investment with the rest of Asia, trade and investment relations with Japan have remained the principle factor driving Thailand's economic development, and this is even more the case because of Japan's position at the top of the Asian regional production hierarchy (Bernard and Ravenhill 1995).

Japanese imports to Thailand include parts for assembly operations, and the overall trade and investment pattern suggests that Japanese investment in the Thai economy involves a significant amount of production for export to the United States (Falkus 1995; Hook 2001). Indeed, the relative importance of Japan's market as a destination for Thai exports has declined slightly since the 1970s, and during 1990–5 it accounted for only 17.2 per cent of Thailand's exports, while the US share of Thailand's exports increased substantially during the 1980s and 1990s, reaching 20.7 per cent during 1990–5 (UN, *International Trade Statistics Yearbook* 1965–97). This pattern illustrates what Rob Steven identifies as Japanese capitalists' 'zonal strategy', designed to overcome the difficulties caused for the country's export-oriented economy by rising domestic wages and trade wars with the United States (Steven 1996; cf. Hook 2001).[25]

As a component of this zonal strategy, Japan has also dramatically increased its ODA to Thailand since 1980. In 1980, total bilateral aid from the Japanese government to Thailand, including grant-aid and yen loans,

[25] Steven notes that one estimate places the amount of Japanese trade with Southeast Asia that is actually comprised of transactions between Japanese parent firms and their subsidiaries at 30–40% of the total trade volume (Steven 1996: 174).

amounted to US$195 million, rising to US$293 million in 1987 and US$408 million in 1991 (Steven 1996: 235, 240). Japanese ODA is generally part of a strategy to facilitate Japanese FDI, and as such it often results from Japanese corporations planning projects involving FDI and then encouraging overseas governments to request Japanese ODA as a source of funding (Shuto 1990: 31; Tienchai 1993: 359–61).

In these various forms, then, Japanese capital has become a major force in Thailand's recent economic development, even as the country diversifies economically, exporting more goods to Europe and importing more capital from other Asian NICs such as Taiwan, Hong Kong, and Singapore. Increases in Japanese FDI spurred by the 'high yen recession' coincided with the onset of double-digit Thai GDP growth (1988–90), suggesting that one of Thailand's main successes in the area of state policy has been to effectively position itself as an ally of international capital and a site for low-cost assembly operations (Falkus 1995: 27–31; Steven 1996: 126–31), a project abetted tremendously by its historical success in repressing, fragmenting, and co-opting labour organizations.

The emphasis here on Japanese economic imperialism as a driving force in Thai industrialization is not meant to imply that Thai capitalists are only secondary actors in the Thai political economy—indeed, in quantitative terms, they continue to dominate investment, even though they do not drive the most advanced industrial sectors (Suehiro 1989: 181–9; Steven 1990: 173; 1996: 251). The issue of foreign manufacturing investment can be put into appropriate perspective, however, if one considers the historic strengths and weaknesses of the Thai capitalist class. Thai capital, particularly in the form of state-backed enterprises and Sino-Thai merchant and banking capital, has been able to carve out niches to which other capitalists have not been able to gain easy access (Hewison 1989*a*). At the same time, the technological weakness of Thai manufacturing capital makes alliances necessary if more productive outlets for investment are to be found (Kunio 1988). Thus, the alliance of Thai and TNC capital is not merely a matter of foreign manufacturing capital dominating Thai markets but is rather a matter of a strategic alliance that allows foreign capital access to markets which would otherwise be difficult to tap and Thai capital access to technology and overseas funds which would otherwise be inaccessible.[26]

The ascendancy of these Japanese-Thai alliances and the maturing of industrial capitalism in Thailand provide a context for understanding political developments since the 1980s. After a series of post-coup manœuvres in the late 1970s, the position of Prime Minister was taken over by former military commander and counter-insurgency specialist Prem Tinsulanond. Prem was less beholden to the CIA than his predecessor, Kriangsak, in part because of

[26] Doner notes that Japanese auto manufacturers have been attracted to Southeast Asia not only because it is a relatively low-cost production zone but because it boasts Japan's most important developing country markets (Doner 1991: 77).

the waning of US military interest in Southeast Asia and the subsequent demise of the Thai Communist Party by the early 1980s. Nonetheless, in juggling competing factions he employed US economic advisers to counter more 'nationalist' economic strains of thought emanating from the faction led by former Finance Minister and Deputy Premier Boonchu Rojanasathian, who recommended a development approach modelled after the East Asian NICs and dubbed 'Thailand, Inc.'. Prem pushed Boonchu's Social Action Party out of his ruling coalition in 1981, clearing the way for economic measures favoured by the IMF/World Bank and the US advisory team and leading eventually to a devaluation of the baht in 1983–4, with an attendant structural adjustment programme being implemented (Anek 1992*b*).

Yet Prem did not simply play the role of IMF/World Bank puppet. When the Thai state proposed to develop an ambitious natural gas and petrochemical scheme for the Eastern Seaboard region in the mid-1980s, the IMF and World Bank opposed it, as did various Thai actors such as the Bangkok Bank, but Prem's administration carried the project forward with support from the Japanese Overseas Economic Cooperation Fund, which became the major source of investment (Hewison 1989*b*: 158). Indeed, Walden Bello has argued that the influx of Japanese capital from the late 1980s prevented Thailand from having to carry through the tariff-reduction measures that were part of the structural adjustment package favoured by the IMF and World Bank as a condition for their loans (Bello 1994: 35; Bello, Cunningham, and Poh 1998: 17). Prem was thus able to consolidate an inevitably uneasy yet comparatively stable rapprochement (which was to last eight years) between more and less internationalized fractions of capital, between Japanese and US planners, and between Thai capitalists and the Thai military.

This delicate balance was disrupted by the regime of Chartchai Choonhavan (1988–91), the first to come to power through open elections since the coup of 1976. Chartchai—who, though a former military leader, headed a coalition that represented the ascendancy of the business classes—allowed money-making activities of all sorts to 'run wild', thus breeding an image of unusually blatant corruption. More fatally, Chartchai attempted to take control over specific state enterprises away from the military and distribute these enterprises to friends and allies in the private sector (Surin 1997: 159–63). The challenge to the military was met by the coup d'etát of 1991, though in this instance the power of capital was substantial enough—and the interests of capital and the military bureaucrats divergent enough—that the military was forced to allow the return of formal civilian rule rather quickly, albeit after its murder of hundreds of demonstrators in May 1992 (Pasuk and Baker 1995: 358–61; McCargo 1997: 257–60).

Thus, while the Thai state remains an important player in the capitalist development process, it is possible to see in the 1991 coup and its aftermath signs that capitalists in Thailand are now powerful enough to have transcended the need for a militarized capitalism that potentially presents challenges to their own more direct hegemony. In this context, the 'democratization' of Thailand is likely to mean fundamentally the expansion of capital and the development of competing centres of political and economic power (polyarchy)—particularly centres that compete with older statist interests. This does not necessarily mean decentralization of economic or political power to non-elite or oppressed groups, and thus it may not mean the development of the sort of democracy many hope will flower as the military becomes more marginal in Thai politics (Anek 1993; Chai-anan 1995). Indeed, the prospects for an expansive form of democracy in Thailand seem limited unless those popular organizations the Thai state has historically repressed—i.e. labour and agrarian organizations—can begin to grow and transform Thai society (Hewison and Rodan 1994; Hewison 1997; Connors 1999; 2003*a*, *b*).

The transformation that has occurred in the period of Japanese quasi-hegemony can thus be understood by revisiting and updating Suehiro's tripod structure. While the three main pillars of capital accumulation remain the same as they were during the period of US hegemony (state firms, Sino-Thai business groups, and TNCs), the process is now mediated by a parliamentary state having strong alliances with the Japanese state and the IMF/World Bank. Military leaders therefore play less of a role than before in the alliances between different legs of the tripod. Instead of the military directly exercising power, it increasingly undergirds such power while bourgeois politicians fight—sometimes in militant fashion—for control over spoils within the parliament (Anderson 1990; Ockey 1992).

3.5 The evolution of labour struggles and labour relations under Japanese quasi-hegemony

The shift in Thailand from accumulation under a global regime dominated by US Fordism to accumulation under Japanese quasi-hegemony and a more diffuse global neo-liberalism is of significance to the labour process, labour relations, and labour struggles. I summarize the significance under two broad headings: (1) the antagonism of Japanese corporations and the global neo-liberal order to even limited, state enterprise-based labour movements and the decline of Thai unions in this context, and (2) the diversification of the Thai working class and the re-emergence of social movement-based labour struggles.

3.5.1 Japanese quasi-hegemony, global neo-liberalism, and the decline of state enterprise unions in Thailand

The dramatic expansion of manufacturing activities in the 1980s and 1990s, and particularly the expansion of women's employment in the export sector, has been the most notable imprint of Japanese quasi-hegemony on the Thai economy. This transformation is salient to the contours of development in Thailand in a number of ways. Japanese capitalists and bureaucrats, unlike US officials in the immediate post-war period, have approached investment in Thailand as a more narrowly economic proposition, rather than as a broader geopolitical issue or a matter of promoting anti-communism. This has meant that competitive wages and production costs are central to investment decisions, and there is far less space than there used to be for the Thai state to negotiate favourable terms of investment by playing the anti-communist card. Furthermore, the World Bank and the IMF, with Japanese and Western investors waiting in the wings, have been putting increasing pressure on the Thai state to privatize more of the remaining state enterprises. In spite of its rhetoric, this privatization push has much less to do with efficiency considerations—most of the state enterprises that have remained since the 1960s have been profitable (Bureau of the Budget 1995–7)—than it has to do with opening up new avenues for investment by international capital (Voravidh 1989: 210, 216).

The combined Japanese-ADB and US-IMF-World Bank push to maintain wage discipline and to privatize industry has proven to be antagonistic to the security of the state enterprise unions. On the one hand, as some of the most highly paid workers in the labour force, state enterprise workers are seen not only by many investors but also by elements of the Bangkok urban professional class as being overly privileged (Sungsidh 1991: 3–5; Chokchai 1994: 134–5). Furthermore, state enterprise union leaders have been opposed to privatization plans, and thus are seen by the state and its international backers as obstacles to neo-liberal economic designs (Voravidh 1989: 217). Rather than becoming a model for a growing trade union movement, then, the state enterprise unions have come under attack since the 1980s and have had their status as a 'labour aristocracy' severely undercut. The ultimate attack was carried out by the military coup regime during 1991, which demoted the state enterprise organizations from unions to associations, which cannot legally call strikes (Sungsidh 1991; Hewison and Brown 1994: 506–8; Sungsidh and Kanchada 1994: 248). State enterprise union leader Thanong Pho-arn, who had opposed privatization of state enterprises, disappeared while the military was in power and is feared to have been killed (Sungsidh 1991: 5; US Department of Labor 1991–2: 7, 19; Hewison and Brown 1994: 508; Chrobot 1996: 29). Furthermore, this time around the return to civilian rule (in 1992) did not bring a quick restoration of the unions. State enterprise unions continued to be formally banned until 2000 (Brown, Bundit, and Hewison 2002: 8), with the imposition of IMF prescriptions for economic recovery as part of the 1997 bailout contributing to the difficulty of their rehabilitation. In this context, the fact that labour affairs were finally elevated to ministerial status in 1993 seems a rather hollow victory (US Department of Labor 1993–4: 3; Hewison and Brown 1994: 509; Mathana and Pawadee 1994: 26).

The banning of state enterprise unions helped illustrate how weak the overall union movement in Thailand has been. Before the ban, only 338,000 out of some 6 million workers in the country were unionized. The ban reduced the number of organized workers to 152,000, or a mere 2.5 per cent of the entire workforce (Sungsidh and Kanchada 1996: 248). Thailand has the lowest level of unionization—and the highest degree of fragmentation, as indicated by the number of peak organizations such as union federations (Chrobot 1996; Bundit 1998)—among all the Asian NICs, and indeed one of the lowest levels of unionization in the world (Frenkel 1993: 311).

To be sure, this does not owe totally to the activities of the state. Structural features of the economy such as the high percentage of the labour force in agriculture and in very small-scale industry, the high capital intensity and thus low labour absorption capacity of the large manufacturing operations, and the high mobility of workers, all make union formation very difficult (Sungsidh 1989: 55; Brown and Frenkel 1993: 90; Sungsidh and Kanchada 1994: 214; 1996: 5–6). Furthermore, corporations in the more labour-intensive industries such as garments have consciously attempted to undermine unions by subcontracting work out to small, non-union firms, often located upcountry and employing underage labour (Voravidh 1996). But state agencies have clearly contributed to the weakness of unions—part of their successful attempt to maintain a highly disciplined, low-wage workforce. Moreover, the previously discussed policies of the state towards agriculture and peasant organizing have contributed to the structural pressures on labour unions by ensuring a large supply of poor rural workers potentially available for labour in urban factories, thus making the threat of dismissal for those who choose to organize a very real one. Overall, while it can be said that the Thai state's labour discipline strategies are not without contradictions, they have in fact been successful in maintaining relatively low wages—at least until the 1990s.

The weakness of Thai labour unions in the era of Japanese quasi-hegemony has been reinforced by the shifting contours of international union activism. Here, it is worth briefly noting the structure of the Japanese political economy and labour relations system—a matter of some contention. Some authors see Japan as quintessentially post-Fordist, while others see it as involving various degrees of Fordism and Taylorism (Peck and Miyamichi 1994: 639–40). Authors such as Martin Kenney and Richard Florida have made much of the unique corporatist structure of Japanese labour relations in large industrial firms, highlighting the existence of long-term employment and multi-skilling of workers, along with job features such as teamwork, job rotation, and substantial worker control over production (Kenney and Florida 1988, 1989; cf. Lazonick 1995). By contrast, Tetsuro Kato and Rob Steven (1991) see Japanese labour relations as centred around highly segmented labour markets that result in job insecurity for most workers, with strict managerial control in the more favoured firms (this being abetted by work teams).

The debate need not be resolved here, but it should be noted that the nature of the labour process and the labour relations system within the core hegemon is not completely unimportant to the development of labour relations on the periphery since there is inevitably some transfer of models from core to periphery. In this respect, the decline of US hegemony and the rise of Japanese quasi-hegemony within Thailand is significant because US agencies such as USAID and AAFLI faded in influence before the Thai state had completely implanted US-based, liberal capitalist labour relations processes, while the overseas arm of the recently unified Japanese national labour federation (Rengo), the Japan International Labor Foundation (JILAF), has stepped onto the scene only since 1990 and has promoted a somewhat different set of labour relations principles.[27] Specifically, JILAF has promoted in-house unions that are meant to formally resemble those of the first-tier domestic Japanese labour relations system, but in a context where (overseas) Japanese companies do not provide the most important benefits available to first-tier workers in Japan, such as lifetime employment (Williamson 1994: 121).

The effects of JILAF's efforts have so far been minimal. Indeed, even though Japan has focused enough on labour policy to have assigned an adviser to the Thai Labour Department (Williamson 1994: 140), the influence of Japanese foreign policy on the labour process in Thailand is still generally regarded as less significant than the influence of either the United States, through AAFLI, or Germany, through FES (Williamson 1994: 187). Moreover, JILAF has sidestepped one of the major tasks of labour organizing in the context of the global feminization of manufacturing labour, failing to recruit many women to Japan for its union leadership training programmes (Williamson 1994: 188).

In this context of reduced US influence and only partial emergence of Japan as an influence on Thai state policy, the labour relations system in Thailand has remained not only weakly developed but lacking in broad coherence, with some elements of the working class still (particularly in the state enterprises) involved in federation-based annual wage rounds developed roughly on the model of US labour-management relations,[28] while other elements (particularly those working for large Japanese firms) are involved solely in in-house negotiations, and yet others (the vast majority of workers) are not organized at all (Bundit 1998). Thus, the Thai working class has in a sense been caught short in the development of disciplinary mechanisms for industrial labour, the Thai state having barely begun to implement labour relations based on US models

[27] JILAF receives most of its funding from the Japanese government and must be regarded as essentially a tool of the Japanese state's broader foreign policy objectives (Williamson 1994: 196–200).

[28] The German Social Democrat Party (SDP), has had increasing influence on Thai labour in recent years through the activities of the SDP's Friedrich Ebert Stiftung (FES). Aspects of FES's favoured state–capital–labour negotiations system are now present in Thailand through the formal principle of 'tripartism'. However, since the state is widely regarded as favourable to management, and since labour's position at the bargaining table is weak, it is labour's ability (or inability) to strike, rather than the seat it has at the table with capital and the state, which is crucial to providing it with any real voice in the system.

when internationalization and increased Japanese economic dominance made implementation of these models contradictory to the latest tendencies in the global capital accumulation process.

3.5.2 The re-emergence and transformation of labour militancy in the 1990s

By the 1990s, these problems in the labour relations system had combined with other factors—such as increasing economic disparities—to produce a hardening of worker attitudes toward management, leading to the first resurgence of labour militancy in the post-1976 period. While no one was calling the Thai labour force radical, observers began noting that the incidence of labour disputes, strikes, and lockouts was increasing, rising from 85 labour disputes with 11 strikes and lockouts and some 100,000 workdays lost nationwide during 1989 to 236 labour disputes with 39 strikes or lockouts and 220,000 workdays lost during 1995 (Department of Labour Protection and Welfare 1989; 1995). The Employers' Federation of Thai Trade and Industry was concerned enough about the trend that it used the occasion of a particularly bitter labour demonstration as an opportunity to encourage enactment of laws controlling worker protests.[29] Yet as Sungsidh Piriyarangsan notes, it is precisely the restrictive labour policies the Thai state has enacted in the past that have been responsible for the most militant labour actions, such as riots and arson attacks: 'If management won't accept unions, workers have no way to express their demands and negotiate in a systematic way' (cited in Fairclough 1997: 81).

Thus, the organized labour movement found itself by the mid-1990s at a crossroads, palpably weak in institutional terms and under fire from corporations and the state at the very time when much of the Thai working class was increasingly showing signs of being disenchanted and willing to act. These contradictions led to new expressions of labour activism, reviving to a limited extent the social movement labour activism that had started to develop through the activities of the NSCT during the 1970s.

The partial re-emergence of social movement labour activism in the 1990s needs to be understood in relation to the changes in Thai labour that have occurred in the recent era of industrial transformation. The upsurge of labour militancy in the 1970s was not merely a recapitulation of older struggles, in spite of its roots in basic economic grievances that workers have shared across time and space. First of all, as elsewhere in Northeast and Southeast Asia, rapid industrial development had transformed the labour force and a significant number of women had entered into manufacturing occupations (Deyo 1989; Lim 1993). Consequently, women were very active in the strikes of the period (especially in the private sector), and some even began to play a leading

[29] *The Nation* (Bangkok daily), 22 Dec. 1996: B8.

role in labour-organizing activities, as they have continued to up to the present, especially after the mid-1980s (Porpora, Lim, and Prommas 1989; Hewison and Brown 1994: 501; Pasuk, Sungsidh, and Nualnoi 1996: 98–9). Second, unlike early twentieth-century labour activism, in the labour activism of the 1970s ethnicity seemed for the first time to play an only minor role, with workers across broad swathes of the economy participating and little reference made to specifically ethnically based demands. Third, as already noted, radical student organizations such as the NSCT played a crucial mobilizing role in relation to labour as well as peasant unions and larger political struggles for democracy. Thus, the labour rebellions of the 1970s manifested the changing identity of Thai labour.

These changes have continued to the present, with significant consequences for the specific character of labour struggle. Labour militancy was relatively muted throughout the 1980s, in part because of the aftermath of military repression and the flight of many surviving student and labour leaders to the jungles, in part because of the effective co-optation of surviving labour groups and institutionalization of their struggles. Furthermore, the economic boom that began in the late 1980s temporarily alleviated concerns about economic opportunity and undercut some of the economic basis of labour militancy. Meanwhile, the transformation of the labour force accelerated, with more workers moving out of agriculture and into industry, and with the feminization of labour within key sectors of the economy continuing (Hewison and Rodan 1994; Sungsidh and Kanchada 1996: 246). By the late 1980s, women accounted for 80 per cent of the workers producing the 10 leading export commodities (Bell 1996: 58–9; Bello, Cunningham and Poh 1998: 81).

In the context of this diversification of the manufacturing labour force, the historical weaknesses of Thai labour organizations and the restrictions placed on them by the Thai state have helped generate a phenomenon that is increasingly prevalent within peripheral countries: labour politics are developing to a greater extent through collaboration with other popular organizations and non-governmental organizations, bringing various workers into political alignment with a variety of non-proletarian social groupings such as village-based organizations and environmental groups (Napaporn 2002). In short, the limitations of the conventional labour politics allowed by the Thai state in the era of neo-liberalism may have the effect of pushing labour back in the very direction from which the co-optive 'bread-and-butter' unionism strategy was designed to divert it: a broader social movement politics. This possibility, though only weakly prefigured at present, can be illustrated by a specific set of events that occurred as the Thai economy was beginning to crumble in the late 1990s.

On 25 January 1997 over 7,000 villagers, organized by a national umbrella organization called the Assembly of the Poor (AoP, or *samecha khon jon*), marched on Bangkok for what was to become a 97-day encampment in front of Government House, involving ultimately more than 10,000 demonstrators, and claiming to represent at least some 50,000 Assembly members. The AoP had been organized several years earlier and had waged demonstrations during the previous administrations of Chuan Leekpai and Banharn Silpa-archa, but the 1997 rally was to be its most powerful political statement to date.[30] The AoP organizers, including representatives of various Thai NGOs, came to Bangkok with a list of 123 specific grievances which they demanded the government address. The vast majority of the issues were forest and land conflicts, but one group in the AoP, an organization called the Council of Work and Environment Related Patients' Network of Thailand (WEPT), was comprised of workers who were demanding compensation for injuries they had suffered, decertification of unqualified doctors from the Labour Court's list of medical experts approved to testify in hearings, and stronger occupational safety and health measures (Napaporn 2002: 96).[31]

WEPT members were mainly from the Bangkok area, but there were also AoP participants who had been involved in actions surrounding the health effects on workers of chemicals used in factories at the Northern Region Industrial Estate (NRIE) in Lamphun province.[32] WEPT was composed almost exclusively of women—though it worked closely with the male staff of a Thai labour NGO—and its leader, Somboon Srikamdokkae, became a leading spokesperson for the AoP as a whole. Eventually, WEPT was able to negotiate successfully for its major demands—including gaining a framework agreement to create an independent worker occupational safety and health commission overseeing the worker's compensation fund and promoting better worker health and safety measures (Voravidh 1998; Napaporn 2002: 97).[33]

Some of the details of these negotiations are important for what they reveal about the role of the Thai state in regulating the labour process as well as for what they reveal about the class-relevant struggles that take place within and through the state. The administration of Prime Minister Chaovalit Youngchaiyudh, which wished to appease voters and defuse a political crisis, selected former Leftists who had gone on to work in the Ministry of Health to negotiate with members of WEPT.[34] The negotiators agreed to the demand for worker compensation and decertification of the doctors, and they also agreed to establish ongoing negotiations over the third demand. Aside from being favoured by the administration with a sympathetic negotiating team, WEPT was also aided by the backing of various groups and individuals within the

[30] For an analysis of the AoP, see Baker (2000).

[31] *The Nation*, 24 Jan. 1997: A2; 26 Jan. 1997: A1.

[32] A number of the observations made in this and subsequent paragraphs are based on conversations the author had with members and supporters of the AoP during February and March 1997.

[33] Overall, when the AoP finally dispersed in early May, it had achieved agreements with the Chaovalit administration on about 30% of the issues and had arranged for on-going talks on others. (*The Nation*, 3 May 1997: A2.)

[34] Personal communication, Thai labour activist. On Chaovalit's propensity towards paternalistic populism, see Chai-anan *et al.* (1990: 118–20, 147–8, 179); *Far Eastern Economic Review* 19 June 1986: 40; 4 June 1987: 45.

medical profession and the Labour Department, who supported creation of the commission.

The political jockeying around the proposal for a worker health and safety commission was complex and included the inevitable tactical disagreements between various groups supporting the workers. Matters were also complicated by the dissolution of the Chaovalit government at the end of 1997 and the willingness of its replacement—the government of Chuan Leekpai—to stick stringently to IMF guidelines and limit state expenditures.[35] Finally, worker health issues and jurisdiction over them were further complicated by the promulgation in mid-1998 of a new Labour Protection Act—a move which could be seen as an attempt on the part of some Labour Department officials to co-opt labour activism, but which also placed these officials at odds with the US Chamber of Commerce and other business boosters, who found the new law too restrictive (Chandler and Thong-Ek 1998).[36]

The 1998 act in fact called for strengthened occupational safety and health measures, requiring the Ministry of Labour and Social Welfare to establish a National Workplace Safety, Health and Environment Committee comprising government, private sector, and worker representatives (Brown, Bundit, and Hewison 2002: 32). This, however, did not meet WEPT's demand for involvement in the committee by academics or health specialists and occupation-related patients—a demand it backed by the collection of 50,000 petition signatures, which under the provisions of the 1997 constitution set in motion a mandatory legislative process to create a new Institute for Occupational Health and Safety (Napaporn 2002: 97–8). Thus, the struggle over the state's occupational safety and health bureaucracy continued into the twenty-first century (a matter noted in Chapter 6). But what can be clearly seen from the protests up to 1997–8 is the way in which they mobilized a complex, hybrid, but class-relevant, and geographically extensive network, situated both within and outside the state. Furthermore, the actions took place outside the normal channels of union politics—though they did not eschew negotiations with state officials—and women workers took the lead in most of the agitating. While the gains from any one action such as this are necessarily quite minimal, they nonetheless illustrate the importance of non-elite groups being able to form trans-class and geographically expansive alliances in order to counter the capitalist use of space against place.

[35] *Bangkok Post*, 18 Nov. 1997.

[36] For the opposition to this act by business groups, see e.g. *Bangkok Post*, 23 May 1998; and *The Nation*, 2 June 1998.

3.6 Results of the industrialization process for workers' wages

The process by which industrialization and proletarianization has proceeded in Thailand during the post-war period has placed great limits on the gains workers have received. Industrialization is typically thought to promise higher wages and a better standard of living than can be obtained for the more marginal agrarian labour force—and sometimes it in fact delivers such benefits. But the conditions of proletarianization in a peripheral context, where internationalized sections of the state oppose strong peasant and labour organizations, have not been conducive to the creation of conditions in which workers could realize gains commensurate with their increases in productivity.

Low farm incomes and the disempowerment of peasants resulting from state activities such as the repression of the PFT have created an abundant, if sometimes only latent, surplus rural labour force. This in turn has underpinned maintenance of very low wages in the industrial sector. Low industrial wages have in turn been further reinforced by the repression and co-optation of labour organizations.

Maintenance of low wages in the industrial sector was neither accidental nor incidental but was, rather, an integral part of the development strategy promoted by international capital and carried out by the internationalized state (Chairat 1988: 170). The strategy was firmly backed by the World Bank and the ILO, as well. Although it is sometimes maintained that wage repression is a temporary strategy, necessary for taking the first steps on the industrial development ladder, the World Bank encouraged low wages throughout the 1970s (Grit 1982: 4, 140) and has indeed continued to do so up to the present.

That a low-wage strategy was favoured by leading elements of the state does not by itself account for the success of the strategy—but the strategy was in fact successful. The evidence of this is necessarily somewhat limited and fragmentary for earlier years of the post-war period, but despite disagreements between various sources on the precise magnitudes, a fairly consistent story of exceptionally low wages—and certainly wage growth well below the rates of GDP growth—emerges from nearly all of the accounts. For example, data presented by Sungsidh shows that between 1947 and 1972 the real daily wage rate for unskilled workers throughout the Kingdom dropped from 8 baht to between 6 and 7.5 baht, while the real daily wage for skilled workers stagnated, moving from 10 baht to between 7.5 and 11.27 baht (Sungsidh 1989: 67). Similarly, ILO data on manufacturing wages in Bangkok, when adjusted for inflation, show stagnation throughout the 1960s, with real hourly wages (in 1963 prices) standing at 3.37 baht in 1961 and at only 3.13 baht in 1970, before jumping somewhat to 3.73 baht in 1971 (ILO 1965–72).[37] This stagnation of wages took place alongside substantial growth in both GDP and manufacturing value added (MVA). The results of such disparities between wage and GDP/MVA growth were the high rates of profit for enterprises found in various surveys for the period. For example, a 1972 survey of Bangkok industrial establishments found annual profits per employee ranging from 7,600 baht to 61,400 baht—at a time when the annual salary for an employee making the daily minimum wage and working 300 days per year would have been 3,600 baht (Sungsidh 1989: 79).

After the implementation of the first minimum wage law in 1972, which raised the official daily minimum to 12 baht, there were a series of formal wage hikes, but these wage gains were quickly eroded by inflation. A Department of Labour survey of all types of unskilled workers during 1972 found that workers between the ages of 20 and 29 made average monthly wages of 479.5 baht, slightly more than their average monthly expenditures of 461.5 baht, while workers between the ages of 30 and 39 made 832.42 baht, well below their monthly expenses of 1,125 baht (Morell and Chai-anan 1981: 194). The minimum wage was poorly enforced: the Department of Labour survey found that 23 per cent of unskilled Bangkok workers made less than 10 baht per day while another 38 per cent made 10–15 baht (Sungsidh 1989: 108). This failure to comply with the minimum wage continued into the democratic period, with the Labour Department reporting in late 1974 that 44 per cent of employers surveyed were not paying the minimum wage—which by then stood at 20 baht per day.[38] Under these circumstances it is not surprising that a 1975 survey by Thammasat University lecturers found the average rate of profit in the industrial sector to be 117 per cent per year, the highest rate in Southeast Asia (Morell and Chai-anan 1981: 194–5).

Real wages of Thai workers grew somewhat in the last half of the 1970s, but with economic slowdown of the late 1970s and early 1980s wage growth was stifled. Moreover, surveys by the Department of Labour in 1982 and by independent scholars in 1985 found that 45 to 48 per cent of workers were paid less than the minimum wage (Limqueco, McFarlane, and Odhnoff 1989: 31, 42). Overall, for the period 1975–90, Sungsidh and Kanchada Poonpanich note that the average real minimum wage for the whole country rose by only 1.65 times while GDP rose 3.1 times (Sungsidh and Kanchada 1994: 238).[39] Consequently, the share of wages and salaries in MVA for the entire Kingdom declined from 24.7 per cent in 1975 to 22.7 per cent in 1990, implying that the

[37] The ILO data are based on recorded minimum wage agreements and not on actual wages paid. It is possible that actual wages were higher, but more probable that they were in fact lower for many workers, given the poor enforcement of wages by state regulatory bodies.

[38] The survey also found that 48% of employers were not observing regulations regarding protection of labour (Morell and Chai-anan 1981: 194).

[39] Indeed, these authors argue that the increase in wages was minimal right up through 1995 (Sungsidh and Kanchada 1996: 19). Below, I will present data suggesting the possibility of somewhat greater wage growth since 1990.

share of gross profit in MVA increased from 75.3 per cent to 77.3 per cent (UNIDO 1992: 45; Mingsarn *et al.* 1995: 2–20). This compares with an average profit share of 66 per cent of MVA for 28 developing countries sampled by the United Nations Industrial Development Organization in the late 1970s (UNIDO 1992: 45; Jansen 1997: 34–5). As Table 3.2 shows, the share of profits in MVA was particularly high in firms employing 20 or more employees. Furthermore, as will be discussed later, throughout the 1980s Thailand had the lowest shares of wages and salaries in MVA of any of the Northeast and Southeast Asian NICs. It also worth noting that the profits generated by low-wage labour were not limited to Bangkok-based industries: a rural industry survey in 1984 found rates of return on fixed assets ranging as high as 194.5 per cent (Saroj 1990: 60).[40] Indeed, a 1989 Thailand Development Research Institute (TDRI) study found the average rate of return on net worth to be even higher for rural industries than for those in the greater Bangkok Metropolitan Region—26.0 per cent compared to 17.2 per cent (Saroj 1990: 61).

Table 3.2. Share of wages in manufacturing value-added, by firm size and sector, Bangkok Metropolitan Region (BMR) and rest of country, 1989–1994 (% of MVA)

Place, firm size, sector	1989	1991	1994
BMR firms, 10–19 employees	34.0	25.5	39.9
Food, beverage, tobacco	43.2	34.0	30.1
Textiles, garments, leather	32.8	24.7	55.7
Chemicals, rubber, plastic	24.9	36.8	18.3
Fabricated metal, machinery	38.4	50.1	37.4
BMR firms, 20 or more employees	13.0	6.1	19.8
Food, beverage, tobacco	9.8	9.5	13.2
Textiles, garments, leather	34.7	27.8	53.9
Chemicals, rubber, plastic	9.2	5.4	8.2
Fabricated metal, machinery	14.5	6.0	9.6
Non-BMR firms, 10–19 employees	20.2	31.7	25.4
Food, beverage, tobacco	14.9	23.5	27.9
Textiles, garments, leather	37.5	53.6	24.7
Chemicals, rubber, plastic	25.5	35.0	38.4
Fabricated metal, machinery	49.9	40.5	19.0
Non-BMR firms, 20 or more employees	16.5	10.3	4.5
Food, beverage, tobacco	15.7	16.4	5.9
Textiles, garments, leather	35.4	8.7	2.1
Chemicals, rubber, plastic	8.8	11.2	1.6
Fabricated metal, machinery	22.1	6.5	14.1

Source: NSO, *Industrial Survey* (1989; 1991; 1994).

[40] This was for the fish-packing industry. Other highly profitable industries included smoked rubber (144.3%), fish meal (118.2%), cold storage (84.3%), and canned food (74.5%). See Saroj (1990: 60).

Summing up wage and GDP growth for the entire period 1960 to 1990, Nikom Chandravithun estimates that real wages increased only 1.5 times while GDP increased 9 times (Nikom 1995: 9–10). The consequences of this have been serious for a workforce that had already begun the period in question making less than subsistence wages. In many locations, surveys done at various times have found workers' wages to be well under the amount required for living expenses. For example, a 1986 survey by TDRI found that unskilled urban workers in the North had incomes equal to only 81.1 per cent of expenses (Pradit 1990: 30). A 1988 survey by the Department of Labour's Wage Commission found monthly household incomes in Bangkok, Phuket, Chon Buri, Saraburi, Korat, and Chiang Mai to be less than monthly expenses (Sungsidh and Kanchada 1994: 241). Another Department of Labour study, at the end of 1990, found wages for all industrial workers in the Kingdom to average 4,000 baht per month, while a single worker, without a family, was estimated to need 4,300 baht per month for survival (Nikom 1995: 10). Even in 1996, after real wages had risen substantially, the Arom Pongpangan Foundation (a Bangkok-based labour NGO named after the now deceased FLUT leader) found that production workers averaged 145 baht per day but needed 201 baht per day for what they themselves would consider a tolerable lifestyle.[41]

In short, throughout most of the post-World War II period, the internationalized Thai state has successfully kept wages extremely low relative to GDP growth—both within agriculture and within industry—and has thus successfully fostered capital accumulation and manufacturing growth through wage repression. In other words, the state has very effectively disciplined labour. This has allowed the generation of large industrial fortunes, both domestically and abroad through what could be appropriately called a 'sweatshop' strategy.

The development of Thai industry, then, while unquestionably *not* a story of blocked development is nonetheless a story of deeply polarizing peripheral industrialization in which the benefits of rapid industrial growth accrue to a few and proletarianization is essentially marginalizing, forcing workers to participate in a labour process that is competitively regulated and repressive towards worker organizations. Moreover, because this labour process is tightly linked to global accumulation dynamics and state activities that enforce competitiveness, the prospects for transcending such conditions of 'sweatshop' labour have so far proved quite limited.

As with the conditions bequeathed by agrarian change, we should resist seeing these outcomes of industrialization for workers as being an inevitable feature of a unilinear development path followed by all industrializing countries. For example, in South Korea, the years of industrial boom from 1967 to 1986 saw wages increase only slightly less in real terms than value added

[41] *The Nation*, 22 Sept. 1996: B2.

(Mazumdar 1994: 550). This is not because the South Korean state failed to discipline labour (cf. Deyo 1989; Ogle 1990; Hart-Landsberg 1993). Rather, because the South Korean state did far more than has the Thai state to discipline *capital*—as emphasized by theorists of the developmental state—it helped foster greater technological adaptation and the development of employment opportunities in industries where higher wages could be offset by higher productivity (Amsden 1989). In short, by disciplining capital as well as labour, the South Korean state took more rapid strides towards leaving behind the 'sweatshop' stage of development than did the Thai state, the latter remaining wed to a development strategy based more fundamentally around low wages.

This difference is all the more interesting given the fact that the South Korean state itself is in many ways highly internationalized. Its internationalization, however, took place in a different way and in a different geographical-historical context than did the internationalization of the Thai state, enabling different developmental choices by the alliance of Korean and US elites who oversaw the process—a matter to which I will return in Chapter 5.

3.7 Conclusion

The evolution of the Thai labour relations system is an integral part of the evolution of the internationalized social factory, which facilitates the reproduction of capitalist social relations within an internationalized state. The process of development in Thailand has led to a rapid decomposition of the peasantry since the 1960s and rapid growth of industrial manufacturing based on moving some members of this displaced peasantry into the factory. The transformation from peasant to proletariat does not happen automatically, however. Aside from the training and education of workers, the state is active in ensuring that workers enter into industrial labour under conditions that don't endanger profits and thus new rounds of capitalist investment. At the same time, the accumulation process must ensure the reproduction of workers themselves. The organization of the industrial labour force plays a crucial mediating role in the distribution of the surplus between capital and labour and thus in balancing the conflicted process of social reproduction.

The ways in which the Thai state has performed the function of disciplining labour and ensuring the social reproduction of capitalism speaks to the state's internationalization. The specific labour regulatory mechanisms of the state have been very much a product of the period of US hegemony, modified—and to a great extent weakened—since the end of the 1970s by the rise of global neoliberalism and Japanese quasi-hegemony within Thailand. In all phases of the development of the post-World War II labour relations system, the internationalization of capital has featured centrally in the actions of the state. The state disciplines labour for a very specific role in the process of peripheral capitalist accumulation. The limitations of the development of unions in Thailand have much to do with the specific character of the accumulation process in the second-tier NICs and the requirement that production costs be globally competitive.

The geographical dimensions of the limitations on labour organizations in Thailand are not only a matter of the 'external' constraints on a national process, however. 'Internal' variations in the topography of labour play a crucial role in determining the precise outcomes of accumulation as well as in political struggles over accumulation's contours. I now turn more fully to these 'internal' variations, examining the transformation of the labour process in Northern Thailand as industrial manufacturing began to make greater inroads into the region in the 1980s.

4

Internationalization of the State under Japanese Quasi-Hegemony: Marginalizing Northern Workers, 1980–2000

4.1 Introduction

The internationalization of the Thai economy and the Thai state analysed in the last two chapters was—like all processes of internationalization—highly uneven. The modern Thai state was formed historically through collaboration between the Siamese monarchy, based in Bangkok, and British colonial officials, with Chinese merchants playing an important subsidiary role (Suehiro 1989; Chaiyan 1994; Thongchai 1994). By the early twentieth century, internationalization of capital and the state under this triple alliance had already led to the emergence of the Bangkok-centred political economy and strongly centralized state that has characterized Thailand throughout the past one hundred years (Dixon and Parnwell 1991). Thus, by the time rapid agrarian and industrial transformation began to take hold in the post-World War II period, it did so against a backdrop of already substantial Bangkok primacy and political dominance.

The patterns of internationalization that have developed in the post-World War II period have largely strengthened this primacy and political dominance. Bangkok was the centre of the new triple alliance based on collaboration between military capitalists, Chinese merchants, and the US Cold War state (Suehiro 1989). As Cold War counter-insurgency and development projects proceeded, significant numbers of displaced peasants left agrarian society to seek urban-industrial employment and, as the overwhelmingly dominant centre of industry, Bangkok received a disproportionate share of the rural-to-urban migration stream, with secondary cities remaining small and economically underdeveloped (Tables 4.1 and 4.2; London 1980; 1985). Consequently, the transformation of urban-industrial labour and the labour relations system described in Chapter 3 took place fundamentally in and around Bangkok, which remained the core area of manufacturing growth.

Table 4.1. Bangkok demographic primacy ratio, 1960–2000

Municipality	Population, by year; primacy ratio (P)[a]									
	1960	P	1970	P	1980	P	1990	P	2000	P
Bangkok	1,299,500	—	1,867,297	—	4,697,071	—	5,882,411	—	6,355,144	—
Korat	19,600	66	66,071	28	78,246	60	208,133	28	204,641	31
Hat Yai	35,500	37	47,953	39	93,519	50	142,592	41	187,920	34
Chiang Mai	41,000	32	83,729	22	101,594	46	166,883	35	174,438	36
Khon Kaen	65,600	20	29,431	63	85,863	55	126,059	47	141,202	45

[a] Primacy ratio (P) = population of Bangkok divided by population of secondary municipality.

Source: NSO, *Population and Housing Census* (1960–2000).

Table 4.2. Regional shares of national population and GDP, 1960–2000 (% of national total)

Region	1960 share		1970 share		1981 share		1990 share		2000 share	
	Pop.	GDP	Pop.	GDP	Pop.	GDP	Pop.	GDP	Pop.	GDP
Bangkok	8.1	23.8	8.9	28.7	10.5	35.8	10.8	39.3	10.4	35.2
Central	23.4	29.3	22.0	27.7	21.7	28.2	22.1	29.9	23.3	35.8
North	21.8	15.8	21.8	15.3	20.1	12.6	19.4	10.0	18.8	8.9
Northeast	34.2	17.0	35.0	15.9	35.2	13.5	34.9	11.8	34.2	11.1
South	12.5	14.1	12.4	12.3	12.6	9.8	12.8	9.0	13.3	9.1

Sources: NSO, *Statistical Yearbook* (1978–95); *Population and Housing Census* (1990; 2000); NESDB, *Gross Regional Product Accounts* (1981–2000).

For most of the post-World War II period up to 1985, the BMR's industrial development was centred on low-wage, low value-added products such as textiles, garments, and low end electronics components, and though there were a number of very large firms in these lines, most manufacturers remained very small in scale, this being the case even among investment-constrained exporting firms. Small size was even more the norm with firms in upcountry regions, where manufacturing development was largely very rudimentary and generally centred in industries such as textiles, garments, and food processing (Table 4.3; Department of Labour 1985–6).

US–Thai collaboration in creating a 'sweatshop' regime contributed to this decades-long emphasis on low-wage, low value-added, Bangkok-centred industry. With the profits of BMR firms extraordinarily high and wage growth flat (as indicated in Chapter 3), the incentive for investors to invest heavily in technological upgrading or training of workers was limited. As a number of studies have shown, this has resulted in Thai firms performing rather poorly in improving their technological capacity, productivity, and worker skills (Limqueco, McFarlane, and Odhnoff 1989; Doner and Ramsay 1994; Deyo 1995*a*)—a problem exacerbated by the Thai state's limited commitment to spending on upgrading workers skills and education (Deyo 1995*b*). Moreover, with profits high and low-wage labour readily available in the BMR on the basis of rural-to-urban migration, relatively few investors sought out alternative locations for production in order to reduce wage bills and other production costs. Thus, the success of the Bangkok-based Cold War alliance in capturing peasants and disciplining labour within a 'sweatshop' regime generated a sort of path-dependent 'involution' (Armstrong and McGee 1968). Displaced rural workers, finding limited non-agricultural employment opportunities up-country, migrated to Bangkok to become part of the industrial labour force, their movement to the BMR itself helping to keep wages down by maintaining a large 'reserve army' of labour; and as such migration patterns buoyed the prospects of the 'sweatshop' strategy, investors continued to plough money into low-wage production in the BMR while workers remitted back to the provinces what they could, allowing some marginalized agrarian producers to stay marginally afloat.

The success of the internationalized Thai state, then, has enhanced the already substantial forces at work producing Bangkok's historical demographic and economic primacy. The character of this primacy has begun to shift slightly, however, in the era of Japanese quasi-hegemony. On the one hand, the BMR's industrial structure has evolved rapidly. Heavy industries and higher value-added industries such as automobile parts have expanded tremendously as the combination of 'flexible' production techniques with relatively low-wage labour has enabled Japanese and other firms to move many assembly operations to developing country sites (Doner 1991; Deyo 1995*a*, *b*; Pasuk and Baker 1998). On the other hand, there has been some *deconcentration* of industry—as opposed to decentralization—with both

Table 4.3. Regional industrial structure, 1981, 1996 (% of GRP, by sector)

Sector	BMR		Central		North		Northeast		South	
	1981	1996	1981	1996	1981	1996	1981	1996	1981	1996
Agriculture	3.9	1.9	28.1	10.6	36.9	20.1	33.9	20.6	41.3	38.9
Mining, quarrying	0.0	0.1	1.8	4.4	1.2	5.6	0.8	0.7	2.5	2.1
Manufacturing	36.1	37.4	22.8	44.7	6.7	13.4	6.7	14.1	6.4	7.2
Construction	4.8	5.3	3.1	5.2	5.0	9.5	5.7	10.2	4.6	6.8
Electricity, water supply	2.5	2.5	1.4	4.0	1.0	2.1	0.9	1.9	1.3	2.3
Transport, communication	6.9	11.2	5.4	5.0	5.9	6.7	5.3	5.9	6.1	7.0
Wholesale, retail trade	19.2	16.8	15.3	11.4	17.8	16.8	21.8	20.5	16.2	14.2
BIR[a]	4.2	10.2	1.6	3.7	1.8	5.4	1.4	5.1	1.6	4.9
Ownership of dwellings	3.0	1.8	5.0	2.7	7.5	5.3	7.2	4.4	4.6	2.9
Public admin, defence	3.2	1.7	4.3	2.4	5.4	4.7	6.6	5.1	4.8	3.9
Services	16.1	11.3	11.4	5.7	10.8	10.5	9.6	11.5	10.7	9.8

[a] BIR = banking, insurance, and real estate.

Source: NESDB, *Gross Regional Product Accounts* (1981; 1996).

newer and older industries expanding from the BMR into surrounding provinces, especially north of the BMR towards Ayutthaya and southeast of the BMR towards the Eastern Seaboard (Map 4.1). Meanwhile, there has also been a more modest urban-industrial transformation in upcountry regions, but with a few exceptions—one of which will be described in this chapter—most upcountry industry remains small scale, low wage, low value-added, and centred in many of the same industries that have long prevailed.

To some degree, the socio-spatial unevenness of internationalization and capitalist industrial transformation appears inevitable (Smith 1986), but the social tensions and strains such unevenness creates have also led to numerous efforts by state authorities to counter uneven development with spatial policies that will undercut some of the antagonism to capitalist development expressed by marginalized groups. The intensity of political struggle in Thailand during the 1970s brought such policy concerns to the fore within the Thai state. The state was already being transformed at this time by changing patterns of internationalization and class (-relevant) struggle. The democratic upheaval of 1973 proved to be the beginning of a very long ending to Thai military capitalism, with the importance of the military increasingly being superseded by the power of a combination of domestic capitalists (primarily Sino-Thai banking families) and more democratically inclined bureaucrats, sometimes allied with Japanese and other foreign investors. International financial institutions such as the World Bank and IMF—with the US government still exercising substantial power through these—also continued to play a central role in relation to the Thai state.

Some state programmes that were implemented in this context aimed to improve rural and agricultural incomes, but others aimed to promote decentralization of urban-industrial development processes, with the backing of the World Bank and other funders. The view of the proponents of these approaches was that the injustices of uneven development were fundamentally spatial and sectoral—too much concentration of high-income manufacturing work in and around Bangkok, leading to skewed national income distribution and overly rapid rural-to-urban migration with consequent poor living conditions for recent Bangkok arrivals—and thus could be dealt with by an effective manufacturing decentralization programme.

In this chapter, I examine one case of such decentralization planning: the state's role in promoting manufacturing industrialization in Northern Thailand, and in particular the targeting of Thailand's major secondary city, Chiang Mai, as a 'growth pole'. Such decentralization exercises are typically analysed quite narrowly, being seen as elements of *national* policy designed to address national imbalances in growth. My approach, instead, is to analyse decentralization planning as an aspect of the internationalization of the state and the continuing, uneven expansion of capital. As such, I emphasize the same sorts of factors that have been shown in previous chapters to be crucial to Thailand's industrial transformation, including the transnational collaboration of specific class groupings and struggles by workers and others over the outcomes of this process. My fundamental claim is that the development of manufacturing industry in the North can be seen as a continuation of the internationalist alliance's incorporation and domination of Northern rural producers. This, in turn, has consequences for the effects of development on those rural producers who are incorporated as a newly constituted manufacturing labour force. Furthermore, since an increasing proportion of these new manufacturing workers are women, unequal gender relations are an important aspect of the manufacturing development process.

Promotion of upcountry manufacturing investment has been the responsibility of a number of key agencies within the Thai state, some of which have already been discussed as agents of internationalization. These include various financial bodies, the NESDB, the BoI, and the Industrial Estate Authority of Thailand (IEAT). I briefly examine the roles of these organizations in promoting manufacturing in Northern Thailand, and particularly in the Chiang Mai extended metropolitan area (CMA). Rather than seeing the activities of these agencies as politically neutral exercises in industrial policy, I interpret them as the latest phase of a longer-term process of class (-relevant) struggle. In this context, I examine the political struggles that have gone on in and around the labour regulatory bodies of the state as the successful promotion of manufacturing in Lamphun province (near Chiang Mai city) and elsewhere in the CMA produced problematic consequences for worker health and the local environment. Moreover, I note the paradoxes of manufacturing in the North for the internationalized state's announced goal of using manufacturing decentralization to redress problems of income disparity. More generally, I illustrate how the dynamics of capital accumulation on the periphery become gendered and spatialized, thus suggesting how geography, gender, and class interpenetrate one another in the development and internationalization of Northern Thailand's manufacturing sector.

4.2 The geography and social structure of uneven development

The industrialization process outlined in previous chapters has produced a highly socially and geographically skewed pattern of economic growth. Thailand has developed some of the highest levels of income disparity in the world, comparable to some of the worst cases of income inequality in Latin American countries (Medhi 1996; Voravidh 1996: 4–5), and the disparities have a strong geographical dimension, being particularly great between urban professionals and entrepreneurs in Bangkok and poor farmers in peripheral areas of the country. Table 4.2 illustrates some of the regional dimensions of Thailand's uneven development. By the 1960s, Thailand's income distribution was already heavily skewed towards Bangkok and the surrounding five provinces of the BMR. This unevenness accelerated under the combined impetus of Cold War counter-insurgency/development schemes in rural areas and import-substituting industrial manufacturing development in Bangkok. By 1978, the BMR and its immediately surrounding provinces to the north and southeast of the BMR accounted for a combined total of 52 per cent of Thailand's GDP, in spite of having only 26 per cent of the total population. Average per capita income in the BMR was 2.7 times the national average. The BMR and the Central Region were even more dominant in manufactured output, accounting for 87 per cent of total national MVA as of 1981—the BMR itself accounting for a full 72 per cent of MVA (NESDB, *Gross Regional Product Accounts* 1994).

These kinds of uneven development phenomena have led to a convenient and widely shared interpretation of Thai economic disparities as being based fundamentally in the failure to transfer enough of the labour force out of agriculture and into industrial or service occupations. The interpretation is convenient because it shifts the focus of analysis from class and gender relations as a basis for inequality to issues of unbalanced regional and sectoral growth, which can then be attributed to particular phases of development or specific state policies rather than to the uneven development dynamics of peripheral capitalism. In contrast, I present here some of the evidence for the importance of class-based income disparities while examining the relationship between these disparities and the geographic-sectoral disparities highlighted by mainstream economists.

Throughout the post-World War II period, the Thai state has successfully kept wages depressed relative to GDP growth—both within agriculture and within industry—and has in fact made wage repression central to its accumulation strategy. The state has not, however, done nearly as much to discipline capital; nor has it successfully addressed the spatial consequences of its wage repression strategy, which has sometimes raised the threat of a socio-spatial disequilibrium detrimental to this Bangkok-based state's own political stability. With low incomes for rural producers and even lower incomes for agricultural workers displaced from the land, people from the provinces began migrating to Bangkok in substantial numbers by the 1960s, hoping to maintain agrarian households through remittances from the somewhat higher wages paid in Bangkok industries (Limqueco, McFarlane, and Odhnoff 1989: 44; Porpora, Lim, and Prommas 1989; Sungsidh and Kanchada 1994: 214). This process contributed substantially to the growth of Bangkok's primacy (London 1980; 1985; Porpora and Lim 1987).[1] Yet many migrants only move to Bangkok seasonally, returning to the provinces during harvest or planting season—a strategy that is not surprising given that Bangkok wages, working conditions, and living conditions are scarcely a large inducement to stay

[1] See London (1985) on Bangkok's relationship with the North.

permanently in the city.[2] Thus, circular and seasonal migration have become central features of the Thai development process. This can be seen as engendering a situation of 'functional dualism' akin to that described by Alain de Janvry for Latin America: by keeping some working members in agriculture, households in effect subsidize the low wages paid in industry (De Janvry 1981). Yet as Jamaree Pitackwong notes, the reverse is also true: by having a family member working in some form of industry, many families are able to continue in agrarian pursuits they would otherwise have to leave (Jamaree 1996: 127–8). Indeed, a pattern of 'involution' characterizes the relationship between the two sectors, with less-than-subsistence level individual incomes in each sector allowing households to keep reproducing themselves at the subsistence level. It is also worth noting that unlike the cases described by de Janvry, in Thailand a very large number of the rural-to-urban migrants are women, a fact accounted for on the one hand by the relatively limited resistance of most rural households to young women taking work outside the community and on the other hand by the great demand for their labour in export manufacturing, sex work, and other urban employment venues (Porpora, Lim, and Prommas 1989; Mills 1999).

Sex work is of particular significance in this context, especially because of its enormous economic role in development and the large amount of foreign currency that the sex industry has helped bring into the country (Truong 1990: 163–5, 173, 177; Wathinee and Guest 1994: 34–6). Estimates of the number of women who are working as prostitutes at any given time vary wildly, from 75,000 to 2.8 million (Wathinee and Guest 1994: 29), but most studies agree that a disproportionate number are women who have migrated to Bangkok from poorer rural areas, and an especially disproportionate number are from the North (Pasuk 1982; Sukanya 1988: 120; Wathinee and Guest 1994: 37–8, 58). Various reasons are forwarded for the disproportionate participation of Northern women, including the possibility that Northern families generally attach relatively less stigma to participation in prostitution—at least if the young women are successful—than do families from other parts of Thailand (Pasuk 1982).[3] Whatever the reasons, however, Northern women have migrated in large numbers for such work—and not just to Bangkok, since Chiang Mai has become an important tourist hub with a substantial prostitution industry of its own (Wathinee and Guest 1994: 35).

With much of the Thai labour force circulating to and from Bangkok for work, the capital city has been the source of approximately 70 per cent of all private, municipal employment (Chalongphob 1993: 395). This mobility, and the high turnover rate of labour it brings, discourages unionization and strengthens the position of capital, especially in Bangkok (Sungsidh and

[2] There are many anecdotes about workers from the provinces who decided they preferred their previous levels of poverty to the shabby working conditions in Bangkok factories. See, e.g. Mills (1997: 47–9).

[3] Author interview, Thai NGO worker, 14 May 1997.

Kanchada 1996: 5), but the primacy of Bangkok has posed certain problems and threats for the elites themselves, including the inadequacy of the urban infrastructure and the danger that in-migrants may swell the urban slum population and potentially contribute to social upheavals. Thus, for some time various state agencies have proclaimed a commitment to reducing Bangkok's primacy and fostering more robust growth in the provinces to stem the tide of in-migration. Yet the state has had only limited success in this project—and perhaps only limited commitment, given the relative social stability that seemingly reigned throughout the 1980s. To explain the state's behavior in response to this issue of Bangkok-centrism, I now examine some of the state agencies that have had the greatest effects on the spatial configurations of industrial growth in Thailand.

4.3 The role of the internationalized state in fostering Bangkok primacy

Social unevenness of economic growth in Thailand has been part and parcel of geographic unevenness, manifest in Bangkok's demographic and functional primacy. Such economic dominance by the capital city is not unusual in Third World or post-colonial contexts, but Bangkok's primacy is especially great (Gilbert and Gugler 1992: 36). Bangkok primacy is the result of a variety of historical political economic factors that have helped create and maintain Bangkok's economic advantages over time. Such factors have created a kind of path-dependence, in which previous levels of political and economic dominance attained by Bangkok generate yet greater dominance by attracting more investment, migration, and political attention to the capital. In this historically contextualized and politicized sense, Bangkok's agglomeration economies and 'market' advantages are real and inescapable in the short term. But these 'market' advantages have been created historically (with the assistance of distinctly 'non-market' forces, such as militaries), and are constantly renewed with the assistance of state (and statist) institutions.

The activities of financial institutions are particularly crucial to the development of Bangkok-centrism in the post-war period and can be used to illustrate how the state has abetted geographically uneven development—both by what one could call its 'active passivity' towards the privately dominated banking sector and by its own use of state tools to reinforce the already skewed patterns of private lending.

The state has done little to impose upon bankers, beyond occasionally raising the amount it requires to be lent to agriculture. This amount had reached 20 per cent of the previous year's deposits by the late 1980s, counting requirements for lending to agriculture, agro-industries, and small-scale industries (Saroj 1990: 65). The state has also placed a ceiling on interest rates, standing at 16.5 per cent per annum in 1990, but these were largely eliminated by the early 1990s and even when in effect the rates of interest actually charged have not been substantially below what the bankers claimed would be their preferred interest rates, according to one survey, so it is doubtful that interest rate ceilings have been regarded by lenders as a large imposition (Saroj 1990: 65, 72–4, 133–5; Unger 1998: 96). Moreover, Thai banks have been able to maintain large interest rate spreads (and thus high rates of profitability), particularly after financial liberalization in the 1980s and 1990s (Unger 1998: 106–7).

Any state restrictions on lenders were, in any event, more than offset by the state's protection of existing banks from 1962 until the mid-1990s through the refusal to issue new licences or to allow foreign banks to establish branch banks, thus creating an oligopolistic, Sino-Thai dominated credit market that has been highly profitable and has spawned some of the most powerful Thai capitalists (Hewison 1989*a*: 188; Unger 1998: 93–100). These bankers, moreover, have been full participants, through various formal and informal consultative bodies, in the process of developing the sorts of state-lending policies that are supposed to regulate them (Anek 1994; Unger 1998: 83–108). The Thai state has also facilitated Thai banking activities offshore by promoting the use of the baht throughout the region (Unger 1998: 98).

Most importantly, private capital has been allowed to dominate the credit market and to move according to 'market logic' (in contrast to cases such as South Korea and Taiwan)—that is, to move in the direction of the bidders who can promise the largest and most secure returns. The social and sectoral consequences of this have included a large preference for lending to industry over agriculture (Bank of Thailand, *Annual Report* 1965–2000). It should also be noted that though the leading Thai bankers have been Sino-Thai and have been protected from international competition by the state, the domestic bankers have shown no aversion to facilitating foreign manufacturing development within Thailand. Foreign firms often bring with them the levels of production capacity necessary for the high rates of return that 'market logic' seeks out (Suehiro 1989: 251–3; 281–6). By contrast, it is often difficult for small-scale domestic firms in provincial areas to get funding from commercial banks—or at least to get it at the most competitive interest rates (Saroj 1990: 16–17, 21–2, 36). Over 90 per cent of industrial credits offered by commercial banks during the 1980s went to medium and large-scale enterprises, even though small-scale enterprises employed the largest number of Thai workers (Saroj 1990: 69, 121; Department of Labour Protection and Welfare, *Yearbook of Labour Statistics* 1990; 2000).

In short, private bankers bet on the strong, and this has not only resulted in preference for manufacturing over agriculture and big firms over small firms—but also in a spatial preference for the Bangkok region, in which the majority of large industrial firms reside. The prevalence of private lending to Bangkok can be seen from Figure 4.1, which shows the credit/deposit ratio of private banks. The high ratio for Bangkok compared to the national average implies that deposits from upcountry regions have effectively underwritten credit expansion in Bangkok. It is also worth noting that within regions, private lending bets heavily on the most developed provinces, and, within these on the provincial seats, or *muang* districts (Saroj 1990: 122–41). Figure 4.2 shows the dominance of Chiang Mai province, and more recently Lamphun province, in the competition to obtain private lending within the North.[4] It is also worthy of note that the North as a whole had the greatest discrepancy during the 1980s between the share of industrial credits in commercial bank credits and the share of MVA in GRP (Saroj 1990: 68). In other words, the North has in general been unable to attract industrial financing commensurate with its level of industrial development.

The tendency of private lending to create agglomerations (indeed, multiple agglomerations at multiple scales) has been abetted by the lending of international institutions such as the World Bank and the Asian Development Bank. As Grit Permtanjit shows, Bangkok was the primary beneficiary of foreign loans between 1961 and 1975, garnering almost 25 per cent of these loans (while another 12 per cent went to the surrounding Central region), compared

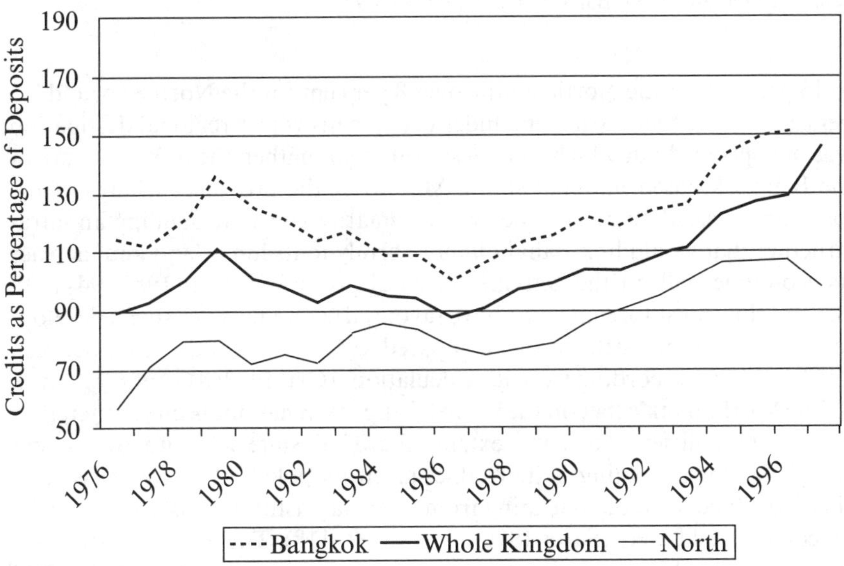

Fig. 4.1. Credit/deposit ratio (annual credit/deposit ratio as recorded in the month of December), commercial banks, 1976–1997.

Source: Bank of Thailand (1976–97).

[4] See Somrudee (1991) on the similarities in the lending pattern to Khon Kaen.

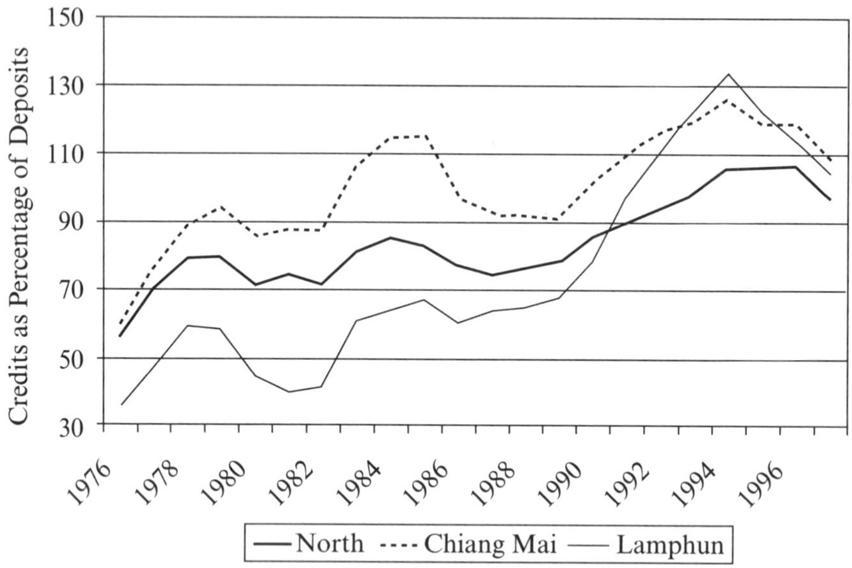

Fig. 4.2. Credit/deposit ratio (annual credit/deposit ratio calculated from statistics for the month of December), commercial banks, Northern Region, 1976–1997.
Source: Calculated from Bank of Thailand (1976–97).

to 15 per cent for the North, a little over 8 per cent for the Northeast, and 7.5 per cent for the South (the remainder being loans whose regional destination was not specified but which were designed to strengthen the infrastructure of the Bangkok-based national state). Moreover, the provinces' shares of the loans are misleading in that they were primarily aimed at building an infrastructure that would link Bangkok more firmly to its hinterland and increase its economic pull on the activities of rural producers (Grit 1982: 94–100). Beyond this, most foreign loans have favoured economic over social development—95 per cent for the former compared to 5 per cent for the latter between 1950 and 1979, according to Grit's calculations (Grit 1982: 84).[5]

The World Bank's 'bet on the winners' approach has not been countered by Thai state policies. To some extent, the Thai state's failure to counter Bangkok-centric lending is inevitable, given its policy of 'active passivity', since 80–90 per cent of all lending from financial institutions is accounted for by commercial banks (Saroj 1990: 6–7, 28, 67, 118). There is only one noteworthy geographic constraint the state places on private lending: the Bank of Thailand requires that commercial banks lend 60 per cent of the previous year's deposits to local groups.[6] Given that banks are not required to open

[5] The pattern is similar from 1980 to the present. See the World Bank's *Annual Report* (1975–2000).
[6] Author interview, Bank of Thailand official, 6 Nov. 1996.

branches in upcountry localities in the first place, their presence there must indicate that the banks' directors find the places in which they have established branches to have some potential, and under these circumstances a requirement to lend a little over half of local deposits to local groups would seem a very minimal constraint.

Beyond this, the activities of lending institutions and facilities created by the Thai state have not countered Bangkok-centrism either. This is particularly evident in the lending of the IFCT. Between 1960 and 1997, 37.6 per cent of the total amount of IFCT lending went to projects in the BMR while another 24.7 per cent went to projects in the Eastern Seaboard and 17.0 per cent to the remainder of the Central Region (IFCT 1960–97). The IFCT, moreover, favours large firms and charges higher interest rates to smaller and riskier firms, as would any commercial bank (Saroj 1990: 22–3, 77, 143).[7] This is accounted for in part by the fact that the IFCT, though founded by the state on the World Bank's recommendation, is in fact underwritten primarily by commercial lenders. So even though it was founded with state capital and continues to draw on such capital it is in fact simply a parastatal that provides opportunities for private entrepreneurs to leverage public (and international) funds, without demanding any clear-cut concessions to broader notions of development or social equity (Saroj 1990: 78). The practices and ideology of the institution have been heavily influenced by the World Bank and USAID and thus the 'bet on the winners' approach which has prevailed in those organizations has also prevailed in the IFCT (Grit 1982: 88–9, 156–7).

Ultimately, whatever the IFCT's policies, a limit is placed on the effectiveness of any of its development-oriented activities by the small proportion of total lending for which it accounts—only 3.1 per cent in 1988, for example. This limitation is even more marked in the case of Small Industry Finance Corporation (SIFC), the only state-assisted industrial lending institution that has dispersed credit somewhat more broadly to upcountry small businesses.[8] Since 1992, SIFC has provided most of its loans to industries outside of Bangkok, including such firms as Northern furniture manufacturers, and it has thus been a potential contributor to the spatial decentralization of industry. Yet SIFC loans have accounted for under 1 per cent of total credit to the manufacturing sector since its creation, and so the relatively greater support provided by SIFC to rural industries has had negligible impact on the overall spatial dispersion of industrial lending (Saroj 1990: 82–3, 118). Thus, the overall tendencies in the social and spatial patterns of industrial lending have been set by the commercial banks and finance companies and, to a much lesser extent, by the IFCT, all of which have bet on the winners and encouraged Bangkok-centric, large-scale industry growth.

[7] IFCT's average interest rates in the 1980s were between 11.5 and 13.5% per annum, somewhat lower than commercial lending rates (Saroj 1990: 63, 65).

[8] Author interview, SIFC executive, 3 Oct. 1996. On the history of SIFC, see, Saroj (1990: 62).

The results have been acceptable to most Thai elites, but they have also led to a number of problems. Beyond the threat of provincial resentment or dissatisfaction on the part of the urban poor leading to rebellion (an occasional occurrence), Bangkok's primacy has led to a number of serious environmental disamenities, including poor air quality, traffic congestion (with high levels of transport fatalities), water and soil contamination, and subsidence with concomitant increases in the problems caused by flooding (Anuchat and Ross 1992; Hussey 1993; Pendakur 1993; Ross and Suwattana 1995). Thus, there has been talk for many years—indeed dating back to the 1960s and 1970s (Muscat 1990: 192; 1994: 97–8)—of the need to encourage greater decentralization of manufacturing growth.

4.4 The tools of provincial industrial development

The Thai state's failure to discipline financial capital in any serious fashion and its more general lack of strong statist controls on industry—reflecting its relationship to specific fractions of capital—has meant that its ability to encourage manufacturing decentralization is based largely on a series of policies that attract investment to given upcountry locations by lowering production costs. These policies have evolved over the years out of various forms of government–private sector interaction, including through the regional chapters of the important Joint Private-Public Consultative Committees (Anek 1992*a*). The specific policies that have been implemented in this context will be discussed directly.

First, however, it should be noted that overall planning for decentralization has been carried out by the NESDB through its 5-year development plans, generally in consultation with World Bank officials. The NESDB has for many years voiced concern over the importance of improving rural incomes and decreasing the disparities between Bangkok and the rest of the country. Decentralization has been argued for on the grounds that it can relieve Bangkok's congestion and pollution while reducing the income gap between the capital city and the rest of the country (Medhi, Pranee, and Suphat 1992: 216–17; Parnwell 1992: 53; Rigg 2003: 226).[9] Moreover, as Jonathan Rigg notes, decentralization may also be attractive to planners because it promises improvement of livelihoods in rural areas without requiring redistributive measures that the state cannot or will not implement (Rigg 2003: 231). The Third National Economic and Social Development Plan (1972–6) called for urban-industrial decentralization through encouragement of industrial relocation. The plan designated nine provincial growth poles, including Chiang

[9] For a more sceptical view of the virtues of promoting decentralization and non-Bangkok-centric growth, see Ayal (1992).

Mai, as loci of the strategy. Implementation began under the Fourth Plan (1977–81) (Medhi, Pranee, and Suphat 1992: 217; Parnwell, 1992: 55).

It can be argued that NESDB plans may have less effect than advertised—even on government policies, let alone on capitalist investment behaviour—because the most the NESDB can do is to mandate certain priorities on the part of ministries. It is not clear that it can enforce performance standards (Rigg 2003: 119–20; Unger 1998: 75).[10] Perhaps more significantly, however, the nature of the NESDB's plans is heavily influenced not only by the training of its officials but by the nature of the groups with whom it confers and collaborates in developing the plans, including World Bank officials. For the most part, this has meant that planning is driven by the needs of major investment groups who can offer the NESDB something in return for its own promises of infrastructure expenditure (Unger 1998: 41).[11] As Michael Parnwell puts it, planning has been growth-first in orientation and has relied on 'the same kinds of urban-industrial centered, top-down strategies which were responsible for enhancing the uneven pattern of development in the first place' (Parnwell 1992: 56).

In any event, the NESDB's framework for luring manufacturing investment away from Bangkok is fundamentally filled out by various concrete measures undertaken by other agencies and by the 'line ministries'. These measures include: (1) the offer of regionally-based investment incentives, such as tax write-offs, by the BoI; (2) the development of productive infrastructure in the provinces through both general state expenditures on transportation and communications and the IEAT's development of provincial industrial estates; and (3) the maintenance by the Ministry of Labour's Wage Commission of a graded minimum wage structure that keeps provincial and non-urban wages lower than those in Bangkok or provincial urban centres.

The BoI's investment zone strategy is perhaps the most widely discussed of these strategies (Ichikawa 1991; Parnwell 1992; Davidson and Ciambella 1995: 243–9; Dixon 1999: 226). The BoI divides Thailand geographically into three such zones, the first being the BMR, the second comprising nearby provinces of the Central Region, and the third covering all outlying upcountry regions. Promoted projects, which must fall into industrial sectors selected by the BoI, are given fewer privileges for locating in Bangkok than for locating further out (BoI, *A Guide to Investing in Thailand* 1993: 34). By providing investors who locate upcountry with the maximum privileges, the stated hope has been to attract more manufacturing projects to the periphery.

Whatever the designs, it is clear that throughout the years of BoI operation, promotional privileges have in fact done little to actually attract investment

[10] However, NESDB officials themselves maintain that these mandates are important directives to the ministries and have significant effects. Author interview, NESDB official, 8 Oct. 1996.

[11] Author interview, NESDB official, 2 Sept. 1996.

away from Bangkok.[12] Projects started under BoI promotion have themselves been highly concentrated in the BMR and the nearby Eastern Seaboard area (Map 4.1; Parnwell 1992: 55), and only very recently have larger numbers begun to appear in outlying regions. It is clear that the promise of tax reductions, the main device available to BoI, is not adequate to compensate for Bangkok's advantages in infrastructure and other necessities.[13]

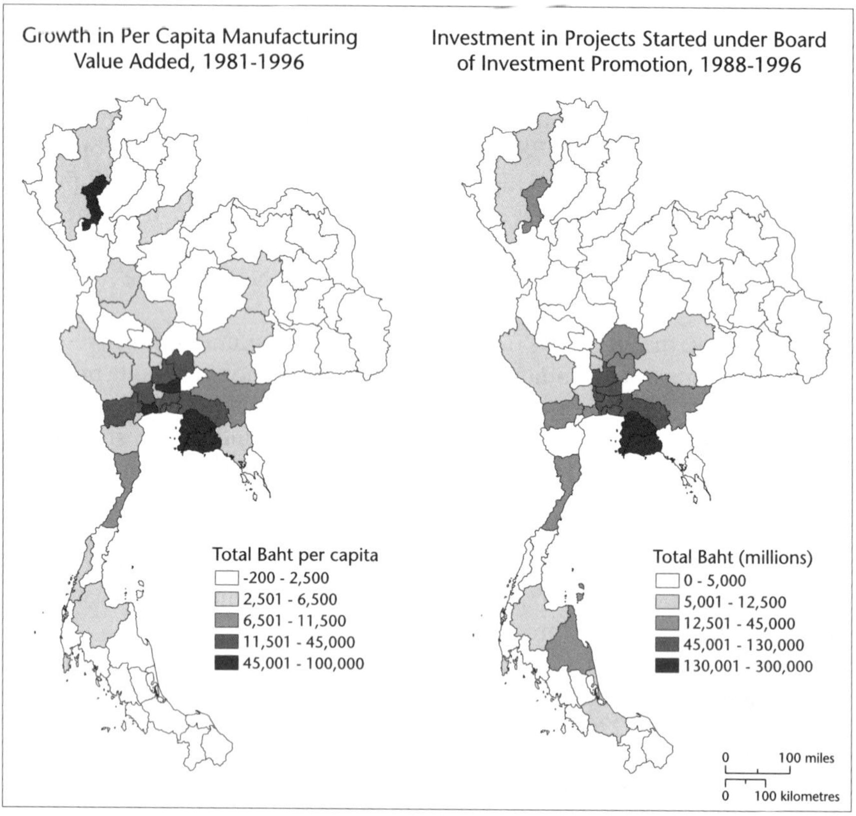

Map 4.1. Thai manufacturing growth, 1981–1996.

Sources: Calculated from NESDB, *Gross Regional Product Accounts* (1981; 1996); BoI, unpublished investment statistics (1988–96).

[12] One World Bank adviser to the NESDB suggested to me, public claims to the contrary notwithstanding, that the actual purpose of the BoI investment zones is *not* to encourage dispersal but rather concentration of industry. He noted the fact that Zone 2, where investors can obtain close to the maximum promotional privileges, is largely adjacent to the BMR, something also true of the majority of industrial estates. Author conversation with Douglas Webster, 20 Feb. 1999.

[13] In any event, a recent study published by the World Bank notes that close to half of all corporations in Thailand declare losses for tax purposes. In 1990, corporate income taxes constituted only 14% of taxes collected by the government, and the tax base as a whole only equalled 20% of gross domestic product, a low level by international standards (Warr and Bhanupong 1996: 75–6).

Given these limitations to BoI policies, perhaps the most important state measures for promoting decentralization have had to do with the development of infrastructure. The effects of infrastructure development have tended to increase rather than decrease centralization, but the building of publicly funded industrial estates and related facilities has been somewhat successful in attracting capital to regions removed from Bangkok—though almost all of these have been within 100–200 kilometres from the capital city, leading to deconcentration and the development of the BMR, rather than decentralization strictly speaking (Parnwell 1992; Bello, Cunningham, and Poh 1998: 117; Dixon 1999: 200, 226). Industrial estates provide companies that locate there with roads, electricity, and amenities such as telephone service, garbage removal, and waste-water treatment, thus reducing production costs and the costs of compliance with environmental regulations. Some also have export-processing zones where manufacturers may obtain more extensive tax exemptions and privileges. Finally, industrial estates allow foreign companies 100 per cent ownership of land, prohibited elsewhere in Thailand, while also allowing firms to bring in foreign technicians and experts without restriction and to take or remit foreign currency (Davidson and Ciambella 1995: 262; IEAT 1995–6: 39).

Up until the mid-1980s, all of the IEAT's industrial estates were located in the BMR or within the Central Region, fairly close to Bangkok.[14] Thus, they had no effect on the development of manufacturing in the poorest and most peripheral regions. Furthermore, they have typically attracted highly capital-intensive industries with relatively low forward and backward economic linkages (Table 4.4).[15] Nonetheless, they have clearly contributed to rising MVA growth rates in the extended BMR throughout the 1970s and 1980s. By the late 1980s, the first industrial estate established outside of Bangkok or the Central region—the Northern Region Industrial Estate (NRIE) in Lamphun—began to have similar effects in the North.

It should be noted that the IEAT, though located within a Ministry of Industry that can exhibit 'nationalist' inclinations (as with the maintenance of various protective tariffs), has been friendly to international capital and to local exporters, since these are some of its most obvious constituents. Indeed, missions to Thailand such as the 1961 Bowen Mission were the first to tout industrial estates, promoting them as a means of encouraging foreign investment (ICA 1961: IX–7–8). However, the IEAT was not established until 1979 (Davidson and Ciambella 1995: 262), when the drive towards export promotion was gearing up and foreign capital was being courted in many of the

[14] By 1996, there were some 25 IEAT-run estates in operation and another 17 planned. There were also some 33 private industrial estates in operation or being planned by this time. Only 3 of the operating IEAT estates were outside of the BMR or the Central Region, though many others were planned (IEAT 1995–6).

[15] For example, a study by Ramstetter (1994) found that foreign-owned multinationals (disproportionately represented in industrial estates), and especially those receiving BoI promotion, had much higher import and export propensities than other firms.

Table 4.4. Average employment and capital intensity of factories, 1992–1996

Promotional agency	Employees per factory	Investment per factory (million baht)
BIA[a]	48	48
PIO[b]	23	18
IEAT[c]	248	615
BoI[d]	324	299

[a] BIA = Bangkok Industrial Authority.
[b] PIO = Provincial Industry Office.
[c] IEAT = Industrial Estate Authority of Thailand.
[d] BoI = Board of Investment.

Sources: BIA (1994); PIO (1992–94); IEAT (1995–96); BoI (1994).

leading export industries (Unger 1998: 78). Industrial estates were thus established less as a matter of domestic spatial policies than as a tool to deepen export-oriented industrialization (EOI), a project that grew as a result of the increasingly important and overlapping interests of foreign investors and domestic exporters. Thus, the IEAT represents another of the more internationalized segments of the Thai state—at least in the orientation of its projects.[16]

If the provision by the state of regional infrastructure is one of the major elements in its manufacturing decentralization project, the role of minimum wage policies cannot be ignored.[17] Until recent years, the minimum wages of different provinces were set as part of a three-tier system, with the BMR having the highest wages, some nearby provinces in the Central Region and popular tourist cities (including Chiang Mai and Phuket) having the second highest wages, and all other provinces having the third highest (see Sungsidh and Kanchada 1996). While competitive wages are not by themselves decisive for attracting manufacturing investment to a particular place, they are nonetheless a necessary condition for attracting such investment (see Narathip 1991).[18] To be sure, a number of commentators on the minimum wage have argued that the regional differences have not been great enough to encourage investment on Thailand's periphery, and recently there have been efforts to create an even more sharply graded minimum wage structure, with minimum wages being

[16] It is worth noting that, as far as the Ministry of Industry more generally is concerned, the main constituents are businesses, often of the national or investment-constrained variety, and these are openly sought out by the ministry for collaboration in programmes such as entrepreneurship training and small-industry development. By contrast, the ministry is not involved in any collaboration with groups such as environmentalists or labour unions. Author interview, Ministry of Industry official, 1 Oct. 1996.

[17] The Bowen mission report, in its discussion of facilities for industry and industry location, cites lower wages as one of the inducements for up-country industrial investment (ICA 1961: IX–8).

[18] Narathip Aramwattanano's study (1991) shows that other important factors include exemptions from business taxes and import duties on machinery and raw materials, freedom to remit profits, and the existence of industrial infrastructure.

determined at the provincial level rather than by broad regionally defined zones.[19] Perhaps of equal importance, however, is the fact that the minimum wage is poorly enforced—and probably even more poorly in upcountry areas than in Bangkok, based on the evidence for the percentages of factories inspected and adherence to labour regulations (Table 4.5). Thus, wages in upcountry regions serve as some inducement to investors not only in the formal sense that minimum wages are lower but in the well-recognized informal sense that paying upcountry workers less than the official minimum is easier. This reality is buttressed substantially by the weakness of upcountry unions and labour organizations. Thus, the specific strategies of the internationalized state's labour regulatory institutions outlined earlier have indirectly facilitated the growth of low-wage, labour-intensive manufacturing on Bangkok's periphery.

4.5 The state development apparatus goes North

I now turn to the case of manufacturing growth in Northern Thailand, focusing particularly on the NRIE and manufacturing development around Chiang Mai. The manufacturing growth which has occurred in the Chiang Mai region is heavily linked to internationalization processes and, as such, is riven with the contradictions and unevenness that mark all capitalist development. This scarcely means that manufacturing development in the North has been 'unsuccessful': indeed, there has been substantial manufacturing growth and in certain respects state policies that have abetted this process can be seen as productive. But the process is not class-neutral, and this means that manufacturing investment is driven primarily by rather narrow capitalist motivations

Table 4.5. Labour Department regulatory activities, by region, 1985–1996 (cumulative totals)

Region	% inspected	% violating regulations	% violators given advice	% violators prosecuted
Bangkok	17.1	48.1	62.0	0.23
Central	13.7	47.5	95.5	0.09
North	11.2	62.4	92.8	0.31
Northeast	11.8	65.0	92.7	0.05
South	10.4	59.3	95.3	0.28

Source: Calculated from Department of Labour Protection and Welfare, *Yearbook of Labour Statistics* (1985–96).

[19] *The Nation*, 29 Aug. 1997; *Bangkok Post*, 10 and 16 Sept. 1997.

rather than by broader social goals. Furthermore, it means that when such social goals are in fact unmet or thwarted by the development process it is difficult to find satisfactory solutions or alternative courses of action within the constraints of the accumulation logic imposed by capital and the capitalist state.

4.5.1 Provincial industry as counter-insurgency

The process of agrarian change displaced a large number of rural producers, set in motion patterns of circular migration, and created a general social upheaval that was intimately connected to the growth of popular movements and the Communist Party of Thailand. While not all migrants to Bangkok were necessarily tinder for labour unions or radical organizations, the social conditions that led to their migration were, and recognizing this, development planners began by at least the 1970s to discuss ways of raising rural incomes through measures beyond agricultural modernization. Indeed, since it was recognized that agricultural modernization necessitated removal of much of the peasantry from the land, there was a concomitant necessity for creation of more employment in the industrial and service sectors. As long as most of the new employment generated in these sectors was centred in Bangkok, migration to Bangkok would be the inevitable solution. But this posed the problems of overburdened Bangkok infrastructure and a skewed regional distribution of wealth, which might undermine state legitimacy both among the residents of growing Bangkok slums and among large segments of the upcountry population.[20]

By 1978, the World Bank was actively supporting industrial decentralization in response to this concern about how to create adequate upcountry employment opportunities. A confidential industrial development strategy report was authored in that year and later published along with the World Bank's policy document on 'participatory' development in Thailand (World Bank 1980; Grit, 1982: 140). The industrial development strategy focused on creating jobs and income opportunities in locations close to pockets of poverty. This was to be done through export-oriented industrial growth, which would rely on concessions to industrial investors, state provision of infrastructure, and the maintenance of a low-wage, non-union work force (Grit 1982: 145–6). The decentralization strategy was in fact simply a component of the broader strategy of attracting export manufacturing investment to Thailand from countries such as Hong Kong, Taiwan, South Korea, and Singapore, where labour costs were rising. Thailand's allure was to be its low minimum wages, standing then at only one-quarter of the average minimum wage in the

[20] On Bangkok slums, see Thorbek (1987); Bello, Cunningham, and Poh (1998: 108); and Dixon (1999: 208–9). Some US planners in Thailand evinced a deep concern about the potential radicalism of recent urban migrants. For example, the Bowen mission report argued that 'It is perhaps better for surplus people [displaced from the countryside by agricultural modernization] to remain on the land near families and food supply than it is for them to become an uprooted urban rabble' (ICA 1961: I–2).

East Asian NICs and only 5 per cent of the minimum wage in industrialized countries (Grit 1982: 140).

The World Bank's industrialization strategy was integrated into the NESDB's Fifth Plan (1982–8). Ultimately, in spite of the World Bank report's rhetorical emphasis on building from the bottom up, the Thai state acted primarily on the suggestion that it promote export-oriented industries spearheaded by foreign investment. Indeed, as the Fifth Plan progressed, the interest of Thai development planners in small-scale industries retreated in favour of larger projects such as the Eastern Seaboard Project (Bello 1994), and by the Sixth Plan (1987–91) the promotion of small-scale and provincial industries had been largely forgotten (Medhi, Pranee, and Suphat 1992: 217).

The neglect of smaller provincial industries by state policy was made possible not only by the strong tendency of capitalist development to favour larger industries but by the collapse of rural insurgency in the 1980s, which made measures designed to legitimize development in the eyes of the poor less urgent. The state, in short, has moved away from promoting provincial industries as a political expedient in the context of counter-insurgency and has instead settled into merely promoting geographic expansion of larger industries as a more general extension of the capital accumulation process. In this context, it has become easy to forget that the earlier designs for provincial industrialization were hatched as part of a broader project for appeasing or subduing unruly peasants. Today, industrial decentralization policies can be presented as politically neutral efforts to spread the benefits of industrialization. Yet the more transparent class character of the project at its inception should not be forgotten since it heavily informs the process of post-insurgency provincial manufacturing development. In the blunt words of a US adviser from the earlier counter-insurgency era, 'Economic development is, after all, one of the best counter-insurgency weapons we have' (cited in Grit 1982: 71). Manufacturing development in Chiang Mai, I thus suggest, has in part been counter-insurgency by other means.[21]

4.5.2 The growth of Northern manufacturing

In spite of Chiang Mai's long history as an urban centre and a regional trade hub—and in spite of various projects to modernize it and build it into a regional growth pole (Chakrit and Hagensick 1973)—manufacturing in the Chiang Mai area before the 1980s remained rudimentary, overwhelmingly small-scale, and centred largely in agriculture. This should not be interpreted as meaning that manufacturing development was entirely stagnant. Indeed, there was some growth in the number of non-agricultural factories over the period from the 1960s to the 1980s, but such factories, while generally being larger-scale than agribusiness operations, tended to have much weaker forward and

[21] For a similar view, see London (1980; 1985).

backward linkages than agribusiness and also had limited economic effects on the Chiang Mai region as a whole (Daranee 1976: 31; Wilawan 1984: 25–7, 41–59; Bhansoon and Vasant 1988: 26, 28).

These points have important and widely recognized implications for development planning that targets poverty and income disparity. Clearly, projects enhancing employment opportunities and growth potential in *dispersed* agro-industries, and which build broader manufacturing projects out of this agro-industry base, will have the most significant employment effects and will spread economic opportunities most equitably, both in a social and geographic sense. Larger-scale manufacturing projects not connected to agriculture are bound to have much less in the way of spread effects and may well contribute to greater social and geographic income disparities within the locale.[22] On the other hand, agro-industries are less likely to lead to rapid economic growth. Thus, continued reliance on them as leading industries within the provinces is likely to result in continued widening of the disparities in wealth between Bangkok and the rest of the country. This point has long been recognized by Thai academics and state agencies (Gray 1990: 46).

In the face of these conundrums, Thai state planners have attempted a multifaceted strategy: promotion of local agro-industry, promotion of certain higher value-added light manufacturing industries, and promotion of tourism, which has significant potential in the Chiang Mai area. The importance of the tourism industry is particularly appreciated by government planners and this has shaped the kinds of manufacturing industries they choose to promote, with lighter industries favoured because they are thought to be less polluting.[23] With regard to small-scale agro-industry or other small businesses, however, it is regularly argued that the state has done less in the way of promotion than most local business leaders feel is necessary.[24]

A recent industrial census carried out by the National Statistics Office (NSO), gives a fair picture of the sorts of changes in the Northern manufacturing base that have occurred since the 1970s and early 1980s, as the state's development policies for the North have taken hold. Whereas agro-industry comprised the majority of manufacturing establishments in the 1970s, by 1996 the largest number of such establishments were in more recently developed industries such as textiles and wearing apparel (Chiang Mai province) and wood products (Lamphun province), with substantial growth in fabricated

[22] Ministry of Industry officials involved in planning for Northern regional development acknowledged to me that they were dissatisfied with the effects of electronics manufacturing on income distribution, but noted that it was promoted because it generated higher growth rates than agro-industry. (Author interview, Ministry of Industry officials, 1 Oct. 1996.) An NESDB official noted that development of Chiang Mai city as a regional light industry hub would increase disparities between the city and rural areas in the rest of the region but argued that this should be acceptable to rural residents as long as their real incomes were rising. (Author interview, NESDB official, 2 Sept. 1996.)

[23] Author interviews, BoI officials, July 1995 and 13 Sept. 1996, and Ministry of Industry official, 10 Oct. 1996.

[24] Author interview, Chiang Mai Chamber of Commerce representative, 25 Oct. 1996.

metal and machinery operations as well. The vast majority of these firms still remained very small scale, with more than 95 per cent of Chiang Mai and Lamphun firms employing less than 10 people, and indeed a significant number of the new manufacturing operations were small firms working as subcontractors for larger ones—over 17 per cent of all proprietors in Chiang Mai and over 34 per cent in Lamphun (NSO 1997).

4.5.2.1 The Northern Region Industrial Estate and the internationalization of capital

In spite of the prevalence of smaller scale, low-technology industries in Northern industrial growth, the growth of selected higher-technology industries in the Chiang Mai region has been significant and has been aggressively promoted by the state. The centrepiece of this strategy since the mid-1980s has been the NRIE, located on 1,788 *rai* of land in Lamphun province. The idea of the NRIE was first broached in the 1970s, construction was begun in April 1983, and the facility was completed in March 1985 (Thongnoi 1983; NRIE 1997). The estate's location, 23 kilometres south of Chiang Mai province's *muang* district along the Chiang Mai–Lampang superhighway and just 2 kilometres east of Lamphun's *muang* district by rail, is strategic in multiple ways (Manat, Udom, and Chukiat 1992: 12; IEAT 1995–6; NRIE 1997). First, the general location of the NRIE, in a region cooler than Bangkok, makes it more attractive to electronics manufacturers, who can reduce necessary temperature control costs by locating in a less hot and humid area.[25] Second the location in a specific area with a nearby urban centre and a large labour force ensures—or so the factories have hoped—an adequate labour supply.[26] Third, the combination of being located near Chiang Mai city but outside of it allows companies to simultaneously take advantage of urban amenities, including skilled workers trained by local universities and technical schools, *and* the lower official minimum wages for non-urban areas. Fourth, the most significant specific advantage of being within the CMA is the ease of access to Chiang Mai international airport, which can be reached by truck in about 30 minutes. For companies shipping goods out of the port in Bangkok, being located along the superhighway that connects with Bangkok (via an approximately 10-hour drive) is also advantageous.[27] Fifth, the specific location within Lamphun province is close to required water supplies (ultimately connected to the Ping River) but is also an area of relatively sandy soils with low agricultural productivity. This led planners to believe that the losses to agricultural production from establishing the estate would be minimal and that local farmers would be willing to sell their land—the latter of which turned out to be true (Manat,

[25] Author interviews, Ministry of Industry officials, 1 Oct. 1996.

[26] Author interview, Ministry of Industry officials, 1 Oct. 1996, NRIE electronics firm manager, 9 Oct. 1996.

[27] Author interviews with NRIE factory managers, 1996–7.

Udom, and Chukiat 1992: 10). Finally, though little-noted in literature on the NRIE, the estate is located in the heart of a region containing some of the most densely populated lands of the Ping River basin, close to Chiang Mai's Saraphi and San Patong districts. These are not only areas that have had significant tenancy and landlessness problems but also areas that were the heart of peasant resistance by organizations such as the Peasant Federation of Thailand (PFT) in the mid-1970s (Bello, Cunningham, and Poh 1998: 154). Thus, the NRIE holds out the prospects of providing industrial employment to nearby households which are among those most affected by agricultural modernization and perhaps the most likely to demonstrate discontent should these employment options not become available.

Despite these locational advantages, however, the NRIE did not originally attract strong investor interest. The estate planners had hoped that not only would electronics firms open operations in the estate but so would a number of more locally oriented firms such as agribusinesses and ceramics manufacturers—the latter being courted because it was thought that the estate could help contain some of the pollution problems which they generate. However, in the early years of the estate there were few firms of any sort that opened projects. Smaller agribusiness firms could not afford land in the estate and in any event did not particularly need the infrastructure and services the estate provided. Likewise, ceramics firms did not find the estate attractive, and electronics firms either did not know of it or had little incentive to invest in Lamphun instead of in Bangkok.[28] Thus, by 1987, there had only been 10 units of land sold in the estate and by 1988 only 3 companies were actually operating (Gray 1990: 47; Manat, Udom, and Chukiat 1992: 12).

The economic boom of 1987–8 changed this. As land prices, wages, and other costs of production rose in Bangkok, investors began to look for alternative production sites. The issue was not so much whether or not to *relocate*—few firms with sunk costs took this sort of option—but rather where to *expand* production or make new investments.[29] In this context, the availability of sufficient infrastructure in Chiang Mai–Lamphun began to pay off, and by 1991 at least 86 units of land were sold (Manat, Udom, and Chukiat 1992: 12). By 31 December 1998, 89 units, representing almost 90 per cent of all units available, had been sold, and 62 of the factories scheduled for these plots were in operation while another 12 were under construction (NRIE 1998).

Of the factories in operation, 21 had set up in the general industrial zone (GIZ) while 41 had set up in the export processing zone (EPZ), which has its own export clearing house to facilitate faster movement of goods through customs for firms engaging solely in export production.[30] The EPZ also provides

[28] Author interviews, Ministry of Industry official, 10 Oct. 1996, and NRIE officials, 22 Jan. 1997 and 29 Jan. 1997.

[29] Author interviews with NRIE factory managers, 1996–7.

[30] An additional 7 bought land in a zone scheduled as commercial and another 4 in a residential zone (NRIE 1998).

additional incentives such as exemption of import duty and value-added tax (VAT) on machinery and raw materials, exemption of export duty and VAT on all exports, and exemption of VAT on local goods used for production (IEAT 1995–6: 38–9; NRIE 1997). At least 46 of the investors in both the GIZ and EPZ received BoI privileges in addition to IEAT privileges (NRIE 1998).

The profile of the types of industries and the nationalities of investors in the NRIE reveals the ways in which Chiang Mai–Lamphun has been brought more fully into the circuits of international capital. Of the 62 factories in operation, 20 were producing or assembling electronics components, 16 were involved in spare parts or machinery, 13 were in the agribusiness sector, and 3 each were invested in jewellery, leatherware, and textiles or wood products.[31] Of the 20 electronics firms, 14 were Japanese owned or were joint ventures involving Japanese owners, while all 6 of the others involved foreign investors from places such as Hong Kong, Switzerland, Singapore, and Korea. Of the 16 spare parts and machinery firms, 10 involved Japanese investors and only 1 was fully Thai owned. By contrast, of the 13 agribusiness firms, 7 were fully Thai owned, while 3 were fully owned by Taiwanese investors (NRIE 1998). In sum, foreign investors dominated the higher-technology production and assembly sectors, while Thai investors were dominant in agribusiness.

Overall, by 31 December 1998, the NRIE was estimated to have received 23,650 million baht worth of investment and to be employing 24,415 persons (NRIE 1998). The electronics and machinery firms were responsible for the greatest amount of employment and invested capital—87.5 per cent of employment and probably over 90 per cent of capital (NRIE 1998).[32] Thus, the NRIE had become a significant outpost of foreign investment in higher-technology export industries with some involvement from Thai agribusiness investors who employed less capital and labour but were still bigger and more capital-intensive than most local agribusiness firms.

4.5.2.2 The 'feminization' of Northern manufacturing labour

A crucial feature of the NRIE's labour force, and that of the manufacturing sector in the North more generally, is the large and increasing number of women employed. Between 1985 and 1999, for example, the number of women in the manufacturing labour force of the Northern Region increased from 207,000 to 353,000, with women outnumbering men by the latter year (NSO, *LFS* 1985–99). This has much to do with the predominance of women workers

[31] Garment and textile factories have largely been excluded from the NRIE because the IEAT does not want to encourage competitors for the local textile manufacturers and exporters of the Chiang Mai area. Author interview, NRIE textile firm manager, 11 Feb. 1997.

[32] Precise figures are impossible to obtain from IEAT publications since not all individual companies list the amounts of invested capital, and the IEAT's listing of NRIE companies for 1995–6 actually shows higher levels of invested capital in the electronics sector alone than that registered in the NRIE's summary of all investment for 1995. Nonetheless, by extrapolating from the percentages given in IEAT (1995–6), as I have done here, one can get a roughly accurate picture of the relative magnitudes.

in electronics firms and other higher-technology enterprises. In the NRIE, women constituted 74.2 per cent of the labour force in 1998 (NRIE 1998). For the region as a whole, a survey of 63 Northern manufacturers (14 from the NRIE) carried out during 1995–7 by Alain Mounier and researchers from ORSTOM found that women constituted 74.8 per cent of all employees in the sampled firms.[33] Women made up 82.5 per cent of electronics employees and 91.3 per cent of textile employees, most of the latter working outside of the NRIE. Women also constituted over 50 per cent of the employees of all construction firms. By contrast, they constituted only 26.6 per cent of the employees of the metal, machinery, and equipment firms (ORSTOM 1997). The figures for the latter two industries demonstrate that gender-based job segregation, though justified typically in terms of 'masculine' and 'feminine' skills and abilities, has more to do with other factors, since construction work involves heavy lifting and activities that are typically taken care of by men within the household division of labour. Exclusion of women from employment in the metal, machinery, and equipment sector, on the other hand, prevents them from having access to some of the highest-paying manufacturing jobs.[34]

Seven NRIE electronics firms at which I conducted interviews—three fully Japanese-owned, one fully Korean-owned, one joint venture with Thai, Hong Kong, and European investors, one joint venture with Japanese and Thai investors, and one joint venture with Singaporean and Thai investors—all virtually exclusively employed women in the production line positions.[35] The rationale given for this was stereotypical and straightforward: women were regarded as having better manual dexterity of the sort needed in the assembly operations and, as one Japanese plant manager put it, 'are easier to control'.[36] Similar sentiments have been expressed by other managers of electronics firms (Theobald, 1995: 27–8).[37]

4.5.2.3 Socio-spatial dimensions of manufacturing development in Chiang Mai–Lamphun

In addition to the feminization of labour occurring as the North industrializes, the spatial dimensions of labour market segmentation and uneven economic

[33] I am greatly indebted to Dr Mounier for sharing ORSTOM's results with me prior to their publication.

[34] Even within construction there is vertical job segregation, with higher-paid positions usually going to men, regardless of whether or not they involve more heavy lifting and other stereotypically masculine tasks.

[35] These 7 firms, at 2 of which I conducted interviews twice, constituted more than a third of the total number of electronics firms in the NRIE as of 2000, when I conducted the last of the interviews. I also interviewed managers of 3 NRIE garment/textile firms, 3 agribusiness firms, and one machinery firm (a carburettor manufacturer), as well as 3 owners or managers of small businesses outside the NRIE in the Chiang Mai area, one a garment factory (discussed below), one a fishing-line factory, and the third an agribusiness.

[36] Author interview, NRIE electronics firms manager, 11 Oct. 1996.

[37] Such sentiments are also heard in industries such as shoes or garments. See e.g. Gray (1990: 193).

growth are worth noting. While increased spatial mobility of labour allows some women from villages to enhance their life prospects through urban migration or daily commuting (Anchalee 1987; Gray 1990), pockets of relative affluence or privilege have developed at multiple scales. The Chiang Mai area as a whole has become increasingly dominant relative to surrounding rural regions, and areas such as Chiang Mai City and the Lamphun NRIE support occupations with relatively high wages compared to even nearby locales.[38] All of these differences have consequences for wages, working conditions, and worker and community solidarity.[39]

It is worth reflecting further on the socio-spatial contradictions that have resulted from the Thai state's successful facilitation of manufacturing growth in Chiang Mai–Lamphun. On the one hand, the increase in manufacturing work and other forms of urban employment has raised overall wage levels. This enabled the North (and the CMA in particular) to more or less maintain its share of GDP between 1992 and 1996, a change from the long-standing trend of relatively declining shares (NESDB, *Gross Regional Product Accounts* 1992–6). In addition, it is possible that there was a slight decrease in aggregate national income disparity between 1992 and 1994 (Kakwani and Medhi 1996). Yet both national-level GRP disparities and national-level income disparities grew during Thailand's growth boom, with the BMR and the Central Region growing much faster than other regions (Table 4.6). Moreover, local

Table 4.6. Percentage increase in per capita GRP, 1978–1996

Region	1978–87	1987–96
BMR	51	94
Central	25	128
North	36	65
Northeast	52	79
South	18	79

Source: NESDB, *Gross Regional Product Accounts* (1978–1996).

[38] Other pockets of manufacturing development have also been evolving in the North, particularly around Lampang and Chiang Rai (see e.g. Flatters and Mingsarn 1994). The NESDB and Ministry of Industry both have various plans for development of more dispersed manufacturing centres throughout the region which will link it with expanding markets and labour in China, labour in Laos, and ports in Burma. (Author interviews, NESDB official, 2 Sept. 1996, and Ministry of Industry officials, 1 Oct. 1996. See also, Amnat (1996).)

[39] There is sometimes differentiation even between neighbouring villages because the major method of labour recruitment for occupations such as construction work is often for workers to go back to their villages to recruit others (Gray 1990: 175–6)—a system also frequently employed in the garment industry. (Author interview, Chiang Mai garment factory manager, 11 Sept. 1996.)

disparities within the North have increased, both between provinces and within localities (Medhi 1996).

Income and manufacturing growth in the upper Northern provinces (Mae Hong Son, Chiang Mai, Chiang Rai, Phayao, Nan, Phrae, Lampang, and Lamphun) has been centred disproportionately in Chiang Mai and Lamphun. Between 1978 and 1996, Chiang Mai and Lamphun increased their combined share of the upper North's GRP from 41.7 to 48.9 per cent, even though their combined share of the upper North's population only increased from 52.5 to 53.6 per cent (NESDB, *National Accounts* 1978; 1996). Manufacturing concentration increased even more dramatically, with Chiang Mai–Lamphun's share of the upper North's total output skyrocketing from 49.6 per cent in 1981 to 87.3 per cent in 1996 (NESDB, *Gross Regional Product Accounts* 1981; 1996). Moreover, manufacturing growth within Chiang Mai and Lamphun—and especially higher value-added manufacturing growth—has been centred overwhelmingly in the CMA, defined here as *amphur muang* Chiang Mai and its seven surrounding *amphur*, along with *amphur muang* Lamphun and its three surrounding *amphur* (Map 4.2; Table 4.7). This suggests that, as with manufacturing growth in the pre-boom period, the manufacturing growth spurred by the economic boom and facilitated by manufacturing decentralization policies has had weak spread effects and has done more to generate intraregional inequality than to generate national level equalization.

4.5.2.4 Gender, class, class-fractional, and occupational stratification of Northern labour

Statistics on wage differentials for different groups of Northern workers indicate the sorts of stratification occurring with industrial growth. For these differentials we can use several sources, including the NSO's *Labour Force Survey* (*LFS*)—which has collected wage data by region, community type, gender, and industry since 1983—and smaller-scale surveys carried out by Varunee Purisinsit and Benja Jirapatpimol, as well by ORSTOM. The smaller-scale surveys provide evidence consistent with the general comparative implications of the NSO data but also supply further details.

LFS data illustrate the success of economic development in raising the overall wages of most Northern workers. Male manufacturing workers in municipal areas saw their real wages increase by 91 per cent, while real wages for female manufacturing workers in municipal areas increased by 132 per cent. Similar real wage increases occurred in non-municipal manufacturing operations, and for women, manufacturing employment was second only to commerce in real wage gains for the period. Yet despite this success, incomes continue to be highly polarized. Overall manufacturing wage levels in the North as of 1997 ranged from 60 per cent of the national average for municipal employees to 75–80 per cent for non-municipal employees. Though improving their relative status, female manufacturing wage-earners still made only 80 per

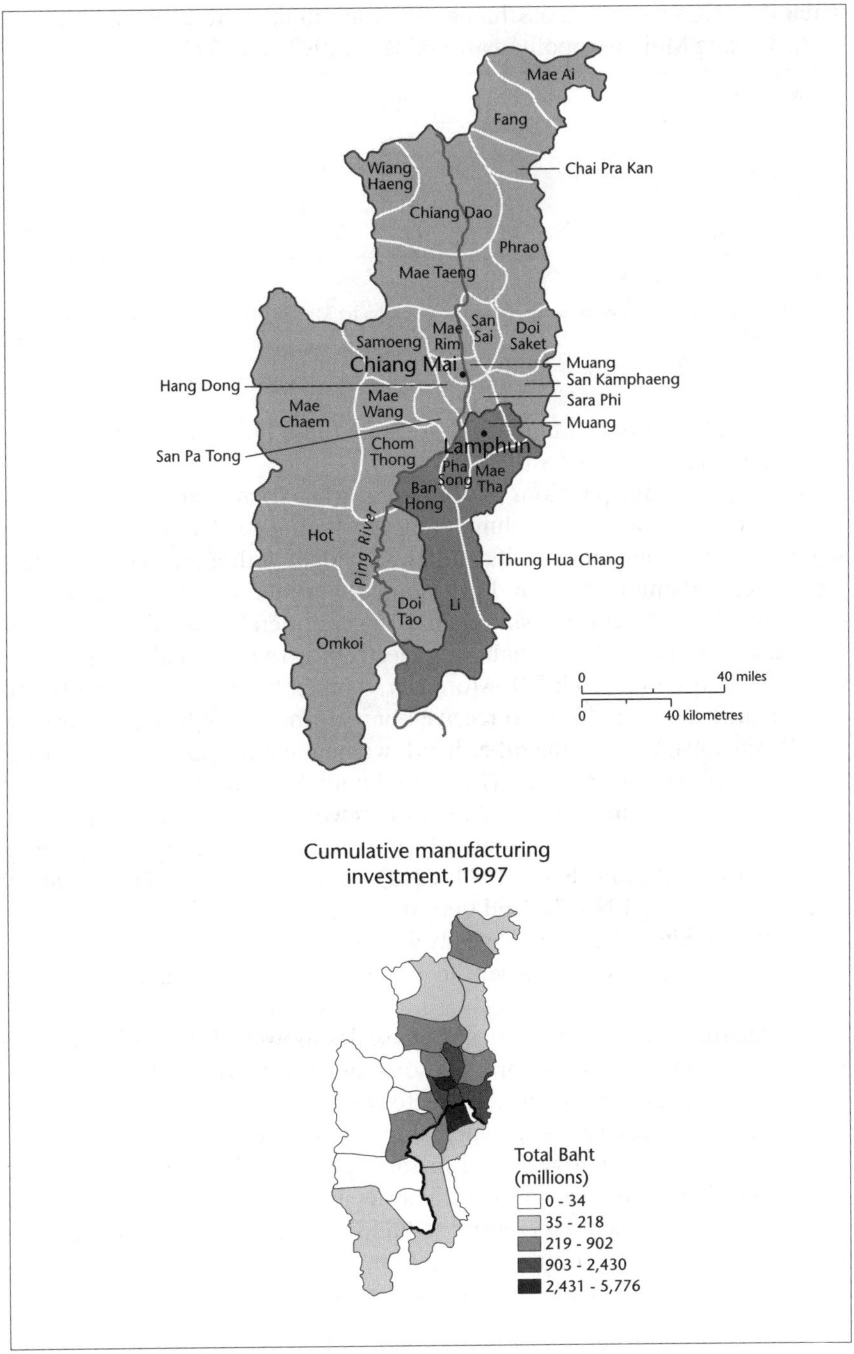

Map 4.2. Manufacturing investment in Chiang Mai–Lamphun, 1997.
Sources: Chiang Mai PIO (1997); Lamphun PIO (1997).

Table 4.7. Concentration of Chiang Mai–Lamphun manufacturing activity in the Chiang Mai metropolitan area (CMA), 1992 and 1997

CMA share of:	1992	1997
Factories	73.0	75.8
Employees	83.6	87.8
Investment	88.4	90.0
Population		53.5[a]

[a] Population figure is for 2000.

Sources: NSO, *Population and Housing Census* (2000); Chiang Mai PIO (1992; 1997); Lamphun PIO (1992; 1997).

cent of what men did in municipal areas and less than 70 per cent in non-municipal areas (NSO, *LFS* 1983–97).[40]

While the relative positions of the lowest wage earners in the North thus improved somewhat without eliminating substantial gender- and place-based segmentation, significant stratification was evident within the manufacturing workforce and among women. For example, Varunee and Benja's data from 1993 show food-processing workers making considerably less (84–5 baht per day) than garment (100–1 baht) or electronics (99–100 baht) employees (Varunee and Benja 1994: 84). Moreover, women working in garments and electronics were more likely to receive bonuses and fringe benefits (Varunee and Benja 1994: 85). On the other hand, women in electronics and garments typically worked longer hours (10 and 9 hours daily, respectively) than did food workers (8 hours daily), and electronics workers averaged an additional 3 hours of overtime per day, bonuses from which further increased their daily earnings (Varunee and Benja 1994: 81–2). Similarly, the ORSTOM survey undertaken during 1996–7 found that food-processing employees made daily wages of roughly 123 baht, while daily wage earners in the textile sector made 149 baht and daily wage earners in the electronics sector made 154 baht (ORSTOM 1997).

The location of work mediates wage scales as well. For example, while agribusiness employment is normally some of the least remunerative factory work, at least one of the agribusiness factories in the NRIE reported paying up to 200 baht per day (including benefits) for seasonal employees.[41] The NRIE agribusiness firms are larger—and in some cases produce more specialized products—than most agribusiness factories in surrounding rural areas. It should not be surprising, therefore, to find that they pay higher wages than rural factories. In line with this expectation, one woman employed in a food-processing factory in San Patong district, several kilometres from the NRIE,

[40] Wages are calculated here in constant prices, using the *LFS*'s February data for private sector employees only.

[41] Author interview, NRIE agribusiness firm manager, 13 Feb. 1997.

reported that daily wages there were 100 baht—at a time (1997) when the official daily minimum wage was 128 baht.[42]

Beyond this, a number of studies have shown that wages have different effects on those receiving them, depending upon the wage earner's precise position in the household and the broader social structure. In general, it is well documented that female labourers in Thailand remit large amounts of money home to support their families (Porpora, Lim, and Prommas 1989: Mingsarn *et al.* 1995: 34). The amount of money women employees get to keep depends also on their occupation, which is in turn related to their age and position in the household (Gray 1990: 196; Varunee and Benja 1994: 66, 71–3).[43]

The picture that emerges then is of a feminized workforce stratified through increasingly complex gradations. Those women who because of either class or class-fractional advantages are able to make it into better-paying urban occupations have gained ground relative to other wage earners, while manufacturing workers in general have gained ground relative to agricultural labour. Within the manufacturing sector, women have made some gains but still earn far less than men overall. Moreover, women working in manufacturing are themselves being increasingly stratified between, on the one hand, those who are younger, single, more educated and able to work not only to support the family but to increase their own consumption, and on the other hand, those who are older, married, less educated, required to support family (including children) and thus more likely to have to endure both lower wages at work and the 'double-shift' at home. It is for reasons such as these that assessments of the benefits and debits of industrial development is a more complex issue than what might be indicated even by attempts to disaggregate its effects along class and gender lines. It is certainly fair to say, nonetheless, that manufacturing employment and industrialization have not presented equally beneficial prospects across the board—a claim that will be reinforced by the discussion of health issues below.

In a fundamental sense, however, the most telling evidence of the social inequities of economic growth comes not from the more fine-tuned stratification occurring *within* the working class but rather from the much more obvious disparities that have developed between capitalists and workers. These can be examined in a number of ways, starting with some of the aggregate regional data available in the NSO's *LFS*. As Table 4.8 illustrates, manufacturing wages in the North have been rising faster than elsewhere in the country, but they have also been rising far less rapidly than manufacturing labour productivity. Such weak wage growth relative to labour productivity growth may be

[42] I stumbled across this information during an informal conversation. While I cannot therefore speak to the generality of these wage levels, it seems certain that factories in areas somewhat removed from major pockets of industrial growth are paying wages well below those prevailing in the NRIE and indeed well below the official minimum wage.

[43] Gray (1990: 196) also found that becoming factory workers allowed women from the village she studied to escape from some, but not all, household chores.

Table 4.8. Changes in manufacturing wages (W) and manufacturing labour productivity (LP) indices, by region, 1983–1999 (1988=100)

Year	Bangkok		Central		North		Northeast		South	
	W	LP	W	LP	W	LP	W	LP	W	LP
1983	88	79	92	91	84	61	102	69	115	93
1984	91	105	95	97	95	100	134	76	97	112
1985	93	89	92	92	108	105	125	80	122	109
1986	94	68	100	103	91	84	127	84	96	101
1987	94	85	83	89	92	111	109	87	88	55
1988	100	100	100	100	100	100	100	100	100	100
1989	91	94	82	96	89	119	102	100	93	62
1990	100	98	89	94	100	126	113	107	95	83
1991	106	100	95	91	110	146	126	97	96	83
1992	115	100	104	97	136	165	106	111	105	105
1993	134	122	120	93	121	140	138	145	102	79
1994	136	118	130	110	144	186	130	131	107	90
1995	120	125	132	122	135	184	145	127	121	102
1996	143	125	143	106	150	204	151	153	133	106
1997	153	105	147	117	164	173	176	171	136	115
1998	140	85	136	107	157	202	164	138	133	96
1999	157	118	143	114	145	229	165	189	137	84

Sources: Calculated from NSO, *Labour Force Survey* (February round, 1983–99); NESDB, *Gross Regional Product Accounts* (1983–99).

due to a variety of factors, but prominent among these is surely the weakness of labour organization in the North compared to other regions of the country.

High rates of labour productivity growth and relatively lower rates of wage growth imply high rates of profit. It is difficult to get direct and systematic evidence for profit rates of given operations within the region, but there is strong evidence supporting the more general thesis that capitalists have done much better during the period of rapid economic growth than have workers. This is suggested, for example, by the income data reported in the NSO's *Household Socio-Economic Survey* (*SES*), which breaks down reported household income by occupation. We can make a very rough estimate of how capitalists and workers did in appropriating the surplus during the boom years by comparing the reported incomes of entrepreneurs and production workers. This comparison inevitably understates the strength of the most important capitalists, since the category of entrepreneur includes both self-employed business proprietors and those employing paid workers. The generally smaller margins of numerous small businesses with no paid workers must pull down the average incomes in the entrepreneur category considerably, yet even so it appears that capitalists fared much better than did industrial workers. For example, between, 1986 and 1994 the average per capita income of entrepreneurs in the Northern region increased from 146 per cent of the national average per capita income to 150 per cent of the average, while the average per capita income of Northern production workers decreased from 120 per cent of the national average to 80 per cent (NSO, *SES* 1986; 1994).

Some of the increased power of capital vis-à-vis labour, it should be noted, derives from specific structural features of the industrialization process in the North. Clearly, the growth of Chiang Mai and the NRIE has helped boost substantially the growth rate of Chiang Mai and Lamphun provinces. At the same time, the industrial transformation of both Chiang Mai and Lamphun on the basis of somewhat more capital-intensive manufacturing has led to a situation where much of the workforce displaced from agriculture cannot find employment in the manufacturing sector and must find work in either construction or various service industries (NSO, *LFS* 1983–99; NESDB, *Gross Regional Product Statistics* 1981–99). For example, in Chiang Mai province, the share of agriculture in total value added declined by over 12 per cent while the share of manufacturing increased almost 11 per cent. Yet the share of the labour force in agriculture declined by 28 per cent while manufacturing's share only increased by 9 per cent, the rest of agriculture's decline being taken up by slightly smaller increases in the construction, commerce, and service sectors. In Lamphun, the imbalance was even greater: manufacturing's share of value-added increased by 50 per cent, but manufacturing's share of the labour force only increased by 16 per cent. Manufacturing wages are not always the highest non-agricultural wages available, but former peasants who cannot find manufacturing labour have much greater difficulty finding higher-paying employment in fields such as trade or clerical work than they do work in fields such as construction or lower-paying field labour jobs. Thus, insofar as the expansion of manufacturing in the high-growth sectors is not labour-intensive enough to absorb a high proportion of the rural producers leaving agriculture, income disparities seem unlikely to improve substantially.

4.5.3 Regulation of labour on the periphery of the periphery

The state's facilitation of capitalist expansion in the North provides a stark contrast with its failure to facilitate extension of labour organizations to the region, and this weakness of Northern labour has prevented it from making strong claims on the surplus it produces. The original focus on creating Bangkok-based and conservative state enterprise unions as the central labour organizations did much in its own right to limit the growth of unions in regions like the North, since development of upcountry unions in the face of local business opposition is bound to be dauntingly difficult without national solidarity (Sungsidh 1989: 122, 129, 185). In addition, destruction of the PFT undermined a potential basis of broader worker solidarity and essentially left the rural producers who were leaving or being pushed out of agriculture to fend for themselves.

Given this background, it is less surprising than it might otherwise seem that through most of its recent history Chiang Mai province—in spite of being home to the country's most prominent secondary urban centre—has had no unions whatsoever. Even less surprisingly, given the complete lack of unions, the province has had very few strikes or work stoppages. Workers do, it will be seen, utilize the 'weapons of the weak' (Scott 1985), through slowing down the work pace or even simply leaving their jobs, but they have been so disinclined to act militantly that throughout the economic boom of the late 1980s there was in fact not a single strike called by any of the few Northern unions (Department of Labour Protection and Welfare, *Yearbook of Labour Statistics* 1986–97).

Some labour unions in Bangkok have attempted to address this problem of low labour militancy in the North by sending representatives to the Chiang Mai–Lamphun area to organize, but the laws requiring that union promoters be a certain minimum age have posed a huge hurdle since most of the large factories that would be the obvious target of unionization campaigns are employing many very young women.[44] Generally, the unions in Bangkok have not been concerned about the prospects of companies moving to places like the NRIE and so have not considered national unionization campaigns to be a necessary defensive strategy.[45] Indeed, one Bangkok-based union organizer who did attempt to encourage union development at the NRIE left after only 3 months, a period some local activists considered insufficient and a reflection

[44] Author interviews, AAFLI official, 8 May 1997, and Thai labour organizer, 13 and 18 May 1997.
[45] Author interview, Thai labour activist, 2 Oct. 1996.

of minimal interest on the part of the unions.[46] Moreover, some critics argue that the labour unions tend to remain narrow and chauvinistic, demonstrating limited concern for immigrant labourers from places like Burma and Laos.[47] Since a large number of workers in Chiang Mai are now being drawn from Burma, this adds to the difficulties of unionization campaigns in the North.

For their part, the international labour organizations also remain limited in their efforts. The AAFLI has undergone some internal changes because of the end of the Cold War, with the undermining of the liberal internationalist labour accord in the United States and what one former AAFLI director calls the end of the 'coincidence of interest' that used to tie US labour leaders to the mainstream of the foreign policy establishment.[48] Yet AAFLI has had limited effect even in Bangkok, and it does little if any training in places such as Chiang Mai–Lamphun. The German labour organization, FES, has organized some meetings in the North, but like AAFLI it is quite limited in what it can reasonably hope to accomplish in an atmosphere of strong state and corporate opposition to labour organizations. Meanwhile, JILAF still has a limited presence, and is mainly active around Bangkok.

Within the Thai state, it is only the Ministry of Labour and Social Welfare that addresses labour issues, and the Labour Ministry is substantially weaker than other state bodies. Moreover, it has no say in development policies affecting labour throughout Thailand.[49] The Ministry's Department of Labour has inadequate personnel to monitor up-country factories. For example, there were only 8 factory inspectors in the Chiang Mai Department of Labour in 1997 and a knowledgeable Labour Department official claimed that a minimum of 50 would be needed to adequately cover all the establishments in the province.[50] The absence of unions interacts with such weaknesses in the Thai state regulatory apparatus, since even under the best of conditions, factory inspectors cannot hope to adequately cover establishments if there are not workers reporting or protesting about conditions—and in the absence of unions, such behaviour by workers is unlikely.[51] The lack of unions to support state regulatory activity, moreover, may not merely be a problem to which the Labour Department has inadvertently fallen victim, since according to critics of the Department it has actually opposed the development of unions in the North.[52]

[46] Author interview, Thai NGO worker, 17 July 1998.

[47] Author interview, Thai NGO worker, 14 May 1997.

[48] Author interview, AAFLI official, 8 May 1997. In 1998, AAFLI and the other AFL-CIO foreign policy institutes were restructured to become the American Center for International Labor Solidarity (ACILS).

[49] Author interview, Thai Department of Labour Protection and Welfare official, 25 Sept. 1996.

[50] Author interview, Thai Department of Labour Protection and Welfare official, 24 Feb. 1997. There were officially 6,000 establishments in Chiang Mai as of 1997 but this official estimates that with various illegal and unregistered operations the actual number is more like 10,000. The staff could, at most, inspect 100 factories per month.

[51] Author interviews, Thai Department of Labour Protection and Welfare officials, 25 Sept. 1996, 24 Feb. 1997, and 7 Mar. 1997.

[52] Author interview, Thai NGO worker, 17 July 1998.

The mutually reinforcing weaknesses of labour unions and state labour regulation have contributed to both the limited claims workers are able to make on the surplus and to dangerous working conditions. The worst of these problems are not found in the large factories of the NRIE but—at least collectively—in the numerous small operations throughout the region that cannot be effectively monitored and have workforces that are too small for successful organizing drives.

The problem is exacerbated by the fact that employment in the 'informal' sector and in jobs such as subcontracting operations have become ubiquitous throughout the region. In addition, a few Bangkok firms have opened small branches in the North to take advantage of somewhat lower wages and (possibly) weaker enforcement. For example, a firm located in Samut Prakan province, near Bangkok, opened a Chiang Mai branch, Northerly Fashion, during 1995, following years of subcontracting work to Northern firms.[53] Northerly employed about 30 women to sew garments for export, primarily to the United States. It was not eligible for BoI privileges, was not located in the NRIE, and still shipped its products out through Bangkok, and thus it appears the parent firm's only purpose in opening the factory was to use it as part of a labour control strategy. Indeed, the Executive Director of the Samut Prakan parent factory indicated that they had opened Northerly in response to rising wages and production costs in Bangkok, and also avoided locating in the NRIE because of reports of labour conflict there—conflicts that will be discussed below. The director also made it clear to me in an interview that the company is strongly opposed to anything but in-house unions and is proud of the fact that its workers are not unionized.[54]

Northerly workers may or may not be paid all that much less—at least relative to their productivity—than Bangkok employees, and indeed the company's management reported displeasure that Northerly was losing money during the first two years of its operation. It attributed this to poor work habits on the part of workers and to their requests for too many holidays.[55] Even so, wages in the factory can hardly be seen as high in real terms. When I visited Northerly during 1996, workers were being paid by the piece and so their daily wages could vary significantly. Workers were guaranteed 80 baht per day, even though when they started they could rarely produce more than 60 baht worth of goods, but this was still well under the official minimum wage. Over time, if their performance improved, they would be guaranteed a minimum of 120 baht per day, still less than the minimum wage. The more experienced women could average around 150 baht per day, and on good days could make as much

[53] Author interviews, Northerly factory manager, 11 Sept. 1996, and Samut Prakan factory manager, 5 Mar. 1997. 'Northerly' is a pseudonym.

[54] Author interview, Samut Prakan factory manager, 5 Mar. 1997.

[55] Ibid. The North has a reputation for having a large number of local holidays, during which most workers expect to be able to return home and participate in social activities, including those Buddhist rituals which allow them to make merit.

as 200 baht. However, it was not clear how many women would be able to perform at this level over time, since the factory had only opened recently.[56]

Northerly scarcely represents the worst of the cases to be found in the Chiang Mai area, and indeed the working conditions inside the factory, at least when I visited, seemed preferable to some of the Bangkok sweatshops that have been the object of concern among labour activists. Nonetheless, the Northerly case illustrates some of the conundrums of late industrialization on the periphery. Even at 120 baht per day, wages are not exorbitant: costs of living in Chiang Mai city are only slightly less than those in Bangkok.[57] Yet such wages are too high to attract industries such as textiles and garments, which must be internationally competitive on the basis of wages. Indeed, so concerned are Japanese textile and garment managers about wages in Thailand that one predicted to me that many of the Japanese investors in this sector would leave Thailand entirely by a few years after the end of the century.[58]

Thus, any very significant wage increases may be, in the internationalized context of Chiang Mai–Lamphun's growth, a potential threat to capital accumulation—or at least to investors' will to remain in the locale. Although profits are unquestionably larger in firms such as the electronics firms of the NRIE, even here wage increases are a disincentive to reinvestment by management, given that virtually all of the production is for export and thus workers' wages are only a cost and not a potential contribution to the firm's market.[59] Firms can recoup their original investments in around 10 years, so there is little incentive for them to stay longer should wage rises continue and lower-wage production sites in places like Vietnam, Laos, or China continue to grow in attractiveness.[60] Indeed, only one of the higher-technology firms at which I conducted interviews, a carburettor manufacturer, was producing substantially for the local market. It thus posed a relatively unusual case of an NRIE firm that was less concerned about wage increases and more interested in long-term strategies for development of its markets—a rational strategy given that it manufactures components used in the motorcycles many of its employees drive to work.[61]

In short, then, the limits of labour organizations, the Thai state, and corporate strategy in the North have all interacted to create conditions in which workers' ability to lay claim to the surplus is weak. To be sure, this has not prevented real wage growth, as noted earlier. The rapid growth of the economy and the mobilization of more poorly paid rural labour for new urban and factory jobs have contributed to high growth rates in wages, a phenomenon

[56] Author interview, Northerly factory manager, 11 Sept. 1996.

[57] Bangkok workers estimated in 1996 that they needed over 200 baht per day to live decent lives. *The Nation*, 22 Sept. 1996: B2.

[58] Author interview, NRIE textile factory manager, 12 Feb. 1997.

[59] Author interviews, NRIE electronics factory managers, 1996–7, 2000.

[60] This estimate of the amount of time needed to recoup investment was provided to me by Alain Mounier in a personal communication.

[61] Author interview, NRIE carburettor factory manager, 3 Feb. 1997.

exacerbated by some shortage in local skilled labour.[62] But the gap between manufacturing wage growth and growth in MVA, noted earlier, is a strong indication that management has found the sort of investment climate it seeks—one which includes a relatively subdued labour force that lays a weak claim to the surplus. Such a labour force, however, was not an inevitability in the North but was the product, as I have shown, of specific social processes and alterable choices. For this reason, the comparative docility of the Northern working class is a phenomenon that can change.

4.5.4 Tragedy and response in Lamphun

The apparent calm of labour relations in the late 1980s and early 1990s gave way to ripples of discontent by 1993–4 when some of the consequences of industrial growth with weak labour organizations and weak state regulation came home to roost at the NRIE (Bello, Cunningham, and Poh 1998: 83–5). Beginning in 1993, at least 10 workers at NRIE electronics firms died from unexplained causes which involved headaches, inflamed stomachs, and vomiting (Forsyth 1994: 35; Theobald 1995: 18).[63] The causes of the deaths have not been definitively determined but both local NGOs and health officials believe that they were probably the result of sustained exposure to heavy metals such as lead. This was a particularly salient problem for the young women working long hours in the electronics factories, who were suffering sustained exposures to chemicals by working overtime in order to make bonus pay (Forsyth 1994: 36).[64]

Local NGOs reacted to the deaths with concern, and related them to environmental problems that had developed in Lamphun since the opening of the NRIE. For example, Lamphun activists had been complaining that waste and untreated water from the NRIE was being dumped into local fields and streams, causing fish kills and other problems.[65] Indeed, the Ministry of Industry's Pollution Control Department acknowledged that the NRIE's incinerator was too small for the volume of waste produced at the estate and had been designed for general waste, not industrial waste.[66] When the worker

[62] This shortage of local workers for higher positions was noted by a number of managers I interviewed. Although the local colleges and universities turned out a fair number of skilled workers, many of these preferred to get jobs in Bangkok, rather than stay in Chiang Mai. (Author interviews, NRIE electronics factory managers, 1996–7.)

[63] The deaths were reported in the Thai English-language press throughout 1994. See e.g. *Bangkok Post*, 24 and 27 Feb., 28 April, and 1 May; and *The Nation*, 27 Feb. and 1 May.

[64] These particular problems are part of a much broader pattern of occupational morbidity and mortality. The leading Thai occupational health specialist, Dr Oraphan Methadilokun, estimates that during 1996 150,000 workers suffered from one form or another of environmental illness, most of this attributable to exposure to toxic substances (Pasuk and Baker 1998: 292). See Glassman (2001) for discussion.

[65] *Bangkok Post*, 1 May 1994.

[66] *Bangkok Post*, 12 Apr. 1994. My interviews at the NRIE suggested that these problems still existed as of 1996–7. (Author interviews, NRIE electronics firm managers, 1996–7.) The manager of a Japanese factory openly told me that his firm considered weak environmental regulations to have been one of the incentives to locate production in Lamphun. (Author interview, NRIE factory manager, 10 Feb. 1997.)

deaths occurred, the two sets of issues galvanized an outpouring of public concern over the effects of the NRIE on Lamphun.

Meanwhile, the workers responded in their own way. None were unionized, and given this and the limitations of state regulation, there were few potential avenues of direct protest. Workers were not involved with NGOs, and most wanted and needed to keep some sort of manufacturing employment, so they could not directly confront management and risk being blacklisted. Instead, when some felt that work conditions were getting too bad, they simply began to quit and seek work elsewhere. A Chiang Mai doctor, who had investigated some of the cases of worker illness, estimated that after the deaths as many as 50 per cent of the workforce at the NRIE may have quit because of health concerns and the fear that they would be next to die.[67] This was in any event a strategy more consistent with the strategies workers at the estate had already learned to utilize. Indeed, despite their reputation for passivity, there had been a number of cases where workers who were not granted wage increases or holidays they asked for quit and went to other firms in the estate, sometimes moving in groups (Theobald 1995).

While the workers' strategy was non-confrontational, there was at least a nascent awareness of the conflictual relationship between workers and management. For example, in a series of interviews and focus group studies with workers from the NRIE, Sally Theobald found that workers consistently complained of long hours, health problems, boredom, and strict rules and regulations, though they regarded these as annoyances which 'we have to put up with to get the money' (Theobald 1995: 30). The women also worried about chemical poisoning at work but found it difficult to get information about the materials they had to work with and noted that 'we're afraid to ask our boss as we'd be seen as trouble makers and he'd say there's nothing wrong anyway' (Theobald 1995: 48). Thus, the relatively meek response of labour cannot be interpreted as complete lack of political consciousness, nor even as complete submission, but should be seen, rather, as reflecting awareness of the extremely weak bargaining position of labour and the need to maintain the income which factory work provides.

A prominent court case involving a woman who worked at the NRIE reinforces the view that Northern workers have good reason to doubt the merits of fighting the system on its own terms. Mayuree Tewiya was employed at the Japanese joint-venture firm, Electro Ceramics.[68] She began working in 1988, at the age of 24, preparing ceramic-alumina plates for soldering before their installation in television sets. In early 1993 she took sick leave from the company after suffering from head and body aches, but she refused to resign unless she received compensation from both Electro Ceramics and the Social Security Office. After a doctor at Chiang Mai's McCormick Hospital diagnosed her as

[67] Author interview, Thai physician, 30 May 1997.

[68] Mayuree's case was discussed in the Thai English-language press. See e.g. *The Nation*, 27 Feb. 1994.

suffering from alumina poisoning, she returned to the company with the diagnosis in hand to demand compensation, but the company refused to accept the diagnosis as valid and eventually ceased to employ her in April of 1994 (Forsyth 1994: 35–6; Theobald 1995: 51). Before going to McCormick, Mayuree had written to Dr Oraphan Methadilokun, the country's leading occupational health specialist, for advice. After the company released Mayuree, Oraphan began helping her prepare a legal case against the company. The case went before the Central Labour Court and was decided in late 1996, with the court ruling that the evidence on the cause of Mayuree's illness was inconclusive.[69]

The case had been billed in advance as a crucial one for both employers and employees since success for Mayuree might open the door for legal claims by other workers.[70] Workers at the NRIE thus could not have missed the significance of the court's ruling for legal fights over health and safety. Yet if Mayuree's case suggests that Northern workers are justified in feeling disempowered, a contemporaneous event suggests they could learn from Bangkok workers that powerlessness is in part the result of failure to organize and act with militancy: the day after the Central Labour Court decided against Mayuree, 200 members of the Assembly of the Poor waged a protest outside the Ministry of Labour in Bangkok and received a 4 million baht compensation payment for 25 workers who were suffering from byssinosis.[71]

4.6 Conclusion

In the Northern region, as elsewhere in Thailand, the internationalized state has contributed to the marginalization of workers. The contradictions of marginalization are embedded in the conflictual dynamics of capitalist internationalization. Workers have been brought onto the margins of capitalism, where they may in some instances enjoy higher wages and more of certain social opportunities than they did previously. Yet at the same time they are made marginal within the very development processes in which they are included, in part because of the weak development of state institutions for disciplining capital or protecting social values not inherently served by capital accumulation. Moreover, while women workers entering the industrial labour force may escape certain types of domestic patriarchal strictures, they also frequently find that they have only been liberated into a realm of life controlled (largely) by male factory managers and owners. Such contradictory and conflictual processes have no predictable outcomes at the level of social struggle: either apathy or explosion is possible at given times and places, depending

[69] *Bangkok Post*, 29 Oct. 1996.
[70] *The Nation* (Bangkok), 9 Oct. 1996: A10.
[71] Ibid. 30 Oct. 1996.

upon vagaries of geography and history which even capital and the state cannot fully master.

More generally, this case study of Northern Thai industrialization shows how class, gender, and the spatial strategies of internationalizing capital and the internationalized state have become intertwined in the regional development process. Regional industrialization, like the industrialization of Bangkok, is not a narrowly internal matter but one in which internationalizing capital expands its geographical range, with the state playing a crucial role in this process. The paradox here is that, while regional industry has been advertised by the state as an answer to inequality, regional manufacturing growth has only mitigated national-level disparities modestly (if at all), while exacerbating disparities within regions and between different social groups and localities. Thus, the manufacturing growth spurred in the North through collaboration of international, national, and local capital has brought both increased productive capacity and economic opportunity and increased *relative* poverty.

It is also worth noting that in some respects the Thai case is different from those of more developmental states like Taiwan and South Korea. The Taiwanese state, as noted in Chapter 2, undertook a land reform that led to a better distribution of agricultural incomes and subsequently to a greater spatial dispersion of industry (Bello and Rosenfeld 1990). The South Korean state, as noted in Chapter 3, has been somewhat more disciplinary towards capital than the Thai state, and although it has its own spatially skewed pattern of urban growth it has been somewhat more successful than Thailand in developing secondary urban centres and in avoiding high levels of primacy (Smith 1996: 121–41; Markusen 1999: 74–9). It thus appears that internationalization and export-oriented industrial development do not inevitably lead towards the *specific forms and levels* of socio-spatial disparity that characterize Thailand, which may instead be the result of geographical and historical contingencies *within* the process of internationalization and uneven development. At the same time, the fact that the Thai economy has grown rapidly and undergone industrial transformation even under such conditions of extreme socio-spatial polarization suggests that a developmental state that disciplines capital and redistributes wealth is not a *necessary* component of the Asian NIC phenomenon that the developmental state paradigm has been designed to help explain. In the next chapter, I will examine this issue further by discussing how the forces of internationalization that have driven Thai development are related to both the broad patterns of industrial growth throughout East Asia and the increase in disparity within Thailand.

5

Interpreting Post-World War II Development in Thailand: More and Less than a National Phenomenon

5.1 Introduction

The processes of internationalization and political economic transformation described in the previous chapters help explain the specific character of recent industrial development in Thailand. Capital accumulation in Thailand has been centred heavily on Bangkok and has favoured a stratum of ruling elites who are disproportionately represented in the capital. The Bangkok-centric political economy has been tightly linked—indeed, over a very long period of time—with broader regional and international processes of capital accumulation, and the Thai elites have been successful at using international connections to buttress their social positions and control. Bangkok elites, in particular, have been able to utilize international support to strengthen a project of Bangkok sub-imperialism, which has in turn brought various local elites from outside Bangkok into national and international coalitions.

All of this has consequences for the results of economic growth and industrial transformation in Thailand. Until the economic meltdown that began in 1996, Thailand's GDP growth record was one of the most impressive in the world since World War II, and the country was included by the World Bank among the 'miracle' economies of East Asia (World Bank 1993), while being lauded by others as 'the Fifth Tiger' (Muscat 1994) and as a new entrant into the ranks of the NICs (Jansen 1991). At the same time, Thailand has become one of the more inegalitarian countries in the world, in terms of income distribution (Medhi 1996; Voravidh 1996) and displays a dramatic spatial skew in the distribution of economic activities. There have also been numerous social and environmental problems connected with industrial development in Thailand, along with various political indignities to the general population (Bello, Cunningham, and Poh 1998)—problems that can be seen alternatively as 'the strains of success' (UNIDO 1992) or as symptoms of 'maldevelopment' (Suthy 1991).

To some extent, each of these images of success and failure correspond to a definite reality of the complex development process, neither of which by itself adequately summarizes the totality. What I focus on in this chapter, however, is not the multifaceted complexity per se but rather the *connections* between what are regarded as the success and failure stories. I argue that facets of development in Thailand that can be seen as contradicting one another in what they imply for development and its prospects are in fact part of a 'contradictory' reality—a reality in which the terms of success in economic growth have been precisely the terms of failure in goals of egalitarianism or empowerment of peasants and workers. The Thai economy has not experienced growing income disparity and socio-ecological problems *in spite* of rapid economic growth; rather it has these problems precisely *because of* rapid growth, and the Thai state has in fact made the very processes that produce income disparity and socio-ecological disamenities the cornerstone of its growth strategy.

The dialectical relations here need to be understood, moreover, in terms that transcend the territorial trap. State power and the political economy in Thailand have been structured in relation to broader regional and international developments, and these developments have much to do with the sort of class power that Thai elites have been able to wield in mastering a highly skewed development process. This process itself forces us to disaggregate the national-level data used to assess industrial development if we are to more adequately comprehend the ways in which internationalization of capital and the state have affected the lives of peasants and workers, men and women, throughout Thailand—as opposed to what these processes have yielded for the mythical 'average' Thai who might be inferred from the national data.

In this chapter, I argue these points by deconstructing the territorial trap in two ways. First, I show that Thai economic growth is less a 'miracle' than it is part of a broader regional phenomenon within the international political economy. Second, I show that the Thai industrialization process can be disaggregated into a class success story for the international alliance that helped bring rapid growth to the country and a much more conflicted process of new opportunities and new dangers for the balance of the population. Throughout this 'deconstruction', I use standard data on Thai economic growth, while acknowledging their weaknesses and limitations. I also argue reconstructively for an interpretation of the data hingeing not only on the numbers themselves but on the kinds of qualitative evidence and theoretical argument I have laid out in previous chapters.

In addition, I discuss the implications of this kind of interpretation of Thai development for studies of industrialization and state policy, assessing in particular what the implications might be for work such as neo-Weberian analyses of East Asian development. I argue that the attention economic geographers have given to issues of scale is important for understanding not only the economic development process in peripheral countries but also the development of state powers on the periphery, which is an integral part of economic and industrial growth.

5.2 The Thai growth 'miracle': more than a national phenomenon

There is little doubt that in terms of GDP growth Thailand presented up until 1996 one of the world's great post-World War II success stories. Economic growth in Thailand has fluctuated according to a variety of factors, but it was positive in all but one post-war year until 1997–8, and in per capita terms Thailand has been among the ten most rapidly growing economies in the world since 1965 (World Bank 1993).

An obvious feature of the East Asian 'miracle' of which Thailand is a part is the geographic concentration of growth success stories: indeed, only one of the top nine GDP performers during 1965–90, Botswana, was not part of either Northeast or Southeast Asia. The reasons for this regional concentration of high-growth economies have been an object of some speculation. A fairly obvious starting point for developing an explanation is an account of the importance of geographical location itself in fostering economic growth. As Peter Dicken puts it, 'To a considerable extent, the formation of . . . geographical concentrations of economic activity can be explained simply in terms of the basic geographical phenomenon of proximity. Firms benefit from being close to their major markets as well as being close to their suppliers' (Dicken 1997: 84).

In the case of Thailand and other East Asian NICs, this matter can be spelled out more explicitly by reference to three intertwined phenomena. First, Thailand and other countries in the region have to a greater or lesser extent been integrated into overseas 'Chinese networks of capital', and this has contributed to shared regional growth phenomena in a variety of ways. It is often suggested, for example, that the dynamism of economic growth in all of the Asian NICs except Japan and South Korea can be attributed in significant part to the performance of regionally extensive Chinese business networks (Jomo 1997: 23). It has certainly been the case, as shown earlier, that the regional connections of Sino-Thai business elites have played an important role in fostering rapid economic growth in Thailand (Pasuk and Baker 1997: 226–8). At the same time, however, it can be argued that stereotyped 'Chinese' business practices, while sometimes stimulating economic growth and helping to maintain a floor under it, can also serve to limit growth. Business relations based strongly on ethnic ties and personal trust may limit entrepreneurial innovation and professionalization of management, while the myriad small Chinese firms that learn over time to survive on small margins may develop a profit-taking outlook that is short-term and inhibits longer-term investment in heavier or higher-technology industries requiring years to mature (Jomo 1997: 25; Kolko 1997: 52). Whatever may be the relative contributions and obstacles to economic growth created by these business practices, though, there can be little doubt that they have in the past created some degree of regional convergence in economic behaviour—including diversification of holdings by leading business families and an orientation towards foreign trade that may have served to limit the extent of import-substitution industrialization strategies (Unger 1998: 55).

Second, an equally important regional phenomenon, particularly in more recent years, has been the ascendancy of Japan as an economic power. Japanese growth has directly triggered economic development throughout both Northeast and Southeast Asia (Bernard and Ravenhill 1995; Hart-Landsberg and Burkett 1998). It is the timing of Japan's yen revaluations that accounts for the timing of the first large wave of Japanese FDI in Southeast Asia during the 1970s (Hook 2001: 43–4) and subsequently, during the late 1980s economic boom in the Southeast Asian NICs (Falkus 1995: 27: Jomo 1997: 27–44). Indeed, more broadly, one can argue that the emergence of a Japanese-led growth dynamic in recent decades can be traced back even further historically to Japanese efforts at constructing a pre-World War II 'co-prosperity sphere' (Jomo 1997: 12–13), and certainly the economic dynamism of both Taiwan and South Korea owes much to conditions which (for better or worse) were created in part by Japanese colonial occupation (Cumings 1984: 8–16; Evans 1987).

A third geographical factor contributing to the concentration of growth 'miracles' in East Asia is the role of the United States in building a regional bulwark against Chinese, Korean, and Vietnamese communism. US aid not only helped rebuild Japan, but provided crucial wherewithal for Taiwan and South Korea at the time of their economic take-offs in the 1950s and 1960s (Cumings 1984: 24; Bello and Rosenfeld 1990: 4; Woo 1991: 45–6), while it also contributed substantially to Thai economic growth. Beyond this, US aid helped stabilize the Northeast Asian political economy and orient it in the general directions favoured by the United States at a crucial period in the development of the region (Wade 1990: 82–3; Woo 1991: 47). Nor was it only direct US aid that was important; rather, access to credit and to the US market was also crucial (Woo 1991: 41; Bello and Rosenfeld 1990: 5).

It is especially worth noting the importance of the geopolitical context in assessing the impact of US intervention on East Asian growth. Given that US concerns were calibrated to longer-term economic considerations, rather than to the short-term interests of US corporations, the assistance the United States provided to anti-communist regimes could be used by states engaged in forms of market intervention that might have been found unacceptable in other contexts (Cumings 1984: 25; Evans 1987: 222; Bello and Rosenfeld 1990: 4; Woo 1991: 47–8). Moreover, the combination of pre-existing class structures and US concerns to undermine communist appeals to landless peasants combined to create a favourable environment for land reform in Northeast Asia—a feature of the development process that has contributed substantially to relatively low income disparities. This differentiates Northeast Asia from Latin America, where the US state has refused to push land reforms that would weaken the position of its strongest political allies (Evans 1987; Amsden 1989: 37–8; Cumings 1990: 471–2; Wade 1990: 83, 241–2; Jenkins 1991: 214–15). It also differentiates the countries of Northeast Asia from Thailand, and perhaps from most other countries in Southeast Asia, as will be discussed further below.

US intervention in Asia also contributed to the regional growth pattern in a particularly perverse way. Japan, South Korea, Taiwan, and Thailand all benefited to various degrees from US war spending. Thus, Japan's post-war economic recovery was triggered by the Korean War (Cumings 1984: 20; Woo 1991: 55), while South Korea, Taiwan, and Thailand all saw rapid growth as a result of the Vietnam War (Cumings 1984: 33; Bello and Rosenfeld 1990: 5; Woo 1991: 94–5). There is consequently a very real sense in which the growth success of the East Asian NICs has been linked to the destruction of economies and societies elsewhere in the region. This is particularly salient for the case of Thailand, whose elites actively cooperated with the US military in the destruction of Vietnam.[1]

Most of these geographic contributions to the regional growth pattern in Northeast and Southeast Asia have been noted in mainstream accounts, though they have been underplayed in many such discussions (e.g. World Bank 1993: 79–81). While recognizing the geographic factors, most analysts of the Asian NICs have focused on specific state policies that are argued to account for why states in the region were able to make so much out of their various geographic advantages. Indeed, much of the debate about the sources of economic growth in East Asia has centred around precisely which sorts of state policies have had the most powerful effect, with the World Bank suggesting limited effects from industrial policy (World Bank 1993: 304–16), various neo-Weberian authors championing it (Amsden 1994; Leftwich 1995; Wade 1996*b*), and certain Southeast Asian institutionalists coming down somewhere in between (Doner and Ramsay 1997; Jomo 1997; Booth 1999). What is not generally examined in detail in these accounts, however, is what has been called into focus in the previous chapters here—namely, the geographical-historical basis of the state that fashions the policies in question.[2] If the nation-state and its development apparatus is seen not as pre-given, fixed, or unitary, a somewhat different account of state policy emerges, one which emphasizes another dimension of the geographical contributions to economic growth.

To view states in this geographically dynamic fashion hardly makes state forms or functions irrelevant. States have indeed played a fairly obvious role in creating sites attractive to investors: specifically, they have facilitated the projects of capitalists in various ways and have repressed labour militancy.

[1] For the effects of the Vietnam War on subsequent economic growth in Vietnam, see Kolko (1997: 20–3).

[2] Some recent work has started to take the geographical-historical specificity of the South Korean developmental state more seriously. See, especially, Woo-Cumings (1995); Cumings (1999*b*); Kohli (1999); and Pempel (1999).

The integration of the Asian NICs into a US-led agenda of anti-communist and pro-capitalist development was one of the most important geographical-historical factors leading to this shared regional orientation among the NICs. Thus, it was not only the positions of the states in question in relation to both communist revolutions and the internationalization of capital that enabled them to experience rapid economic growth; rather it was their active constitution as capitalist states, which adopted pro-US policies, that eventually placed them in the path of friendly capital flows. Constitution of the state was a process that itself evolved throughout the post-war period and led to transformation of state forms and functions—this including recent transformations under the conflicting impulses of various Western-trained (neo-liberal) and Japanese-identifying (neo-statist) state managers. Thus the states that fomented economic growth in Thailand and elsewhere in East Asia did not exist as pre-given and did not simply 'insert' themselves into the international arena but were rather transformed and internationalized as part of the process of capitalist internationalization.

In sum, it is evident that rapid economic growth in Thailand has been more than a national phenomenon and links Thailand to the development of a regional economy whose boom was integrally related to vagaries of the international political economy (Bernard and Ravenhill 1995; Hart-Landsberg and Burkett 1998). Since these vagaries include the explosive uprising of peasants and nationalist political blocs in East Asia, along with US-led resistance to this uprising, Thai economic growth is very much part of a complex, internationalized process of class-relevant struggle. This internationalized class basis of economic growth and state policy in turn helps explain certain aspects of uneven development within Thailand.

5.3 The Thai growth 'miracle': less than a national phenomenon

The internationalist dimensions of both state activity and economic development in Thailand contribute to an explanation of why the economic growth boom has exhibited tremendous socio-spatial unevenness—a feature of Thai development noted by a wide range of observers (see e.g. Parnwell 1996). Thai development has been driven to a substantial degree by the interests and actions of the most internationalized class fractions—i.e. by the internationalized Thai and foreign capitalist elite (Schmidt 1996; Hersh 1997). This elite has no specific reason to be concerned with broad-based or egalitarian development in Thailand per se and has systematically encouraged Bangkok-centric growth, based on repression, exclusion, or marginalization of peasants and workers (Chai-anan 1995; Pasuk, Sungsidh, and Nualnoi 1996; Schmidt 1996). Thus, the socio-spatial inegalitarianism of Thai growth is not accidental or incidental but is integrally related to the growth dynamics of a particular highly internationalized regime.

In this section, I present empirical evidence regarding the inequities in the pattern of development produced by this regime in Thailand. Assessing disparities in economic growth necessarily involves numerous complexities and ambiguities, and thus the quantitative evidence offered here needs to be interpreted cautiously. In this section, I attempt a judicious use of this data to show that Thailand's growth boom has been less than a national phenomenon, having strong socio-spatial biases that marginalize much of the population within the growth process, even where it improves standards of living.

5.3.1 Economic growth and poverty reduction

One of the primary legitimizing arguments made for Thailand's pattern of economic development is that in spite of its recognized problems it has reduced poverty levels and raised living standards for the majority of the population (e.g. Warr 1993; Muscat 1994). However, caution is needed to avoid either, on the one hand, downplaying the real economic gains that have come to much of the Thai population with capitalist development or, on the other hand, overstating the depth and breadth of these achievements.

Where poverty reduction is concerned, the standard statistical evidence normally presented—based on the NSO's *SES*—shows an enormous success story. Though historical evidence regarding peasant food consumption suggests that poverty was not extreme in Thailand prior to World War II (Zimmerman 1931; cf. Dixon 1999: 58), the first systematic study of poverty conducted in 1963 found poverty incidence to be 57 per cent, 61 per cent in rural areas (Suganya and Somchai 1988: 15). Poverty incidence declined to 32 per cent in 1975–6 and to 24 per cent in 1981, though most of the apparent decrease between 1975–6 and 1981 was due to a change in the NSO's definition of households (Medhi 1993: 404–5), and the official incidence of poverty increased again to 29.5 per cent in 1986, the vast majority of those living in poverty being those who worked in the agricultural sector (Suganya and Somchai 1988: 42–3).

The growth boom of the late 1980s and early 1990s reduced official poverty incidence very rapidly. The share of the population living under the poverty line declined to 13.1 per cent in 1992 and then took a further sharp drop to 9.6 per cent in 1994, with only 1.9 per cent of the urban population counted as living in poverty during the latter year, apparently leaving poverty a fundamentally rural phenomenon, and one especially associated with the Northeast (Kakwani and Medhi 1996; Dixon 1999: 220).[3]

[3] It should be noted, however, that cases of urban poverty may be badly under-counted, both because of inadequate sampling (Medhi 1996: 21; Bello, Cunningham, and Poh 1998: 108) and because of a failure to account for a large influx of recent migrant workers to the extended BMR (Dixon 1999: 208–9, 222). Moreover, urban poverty appears more significant when one uses measures other than the standard

This picture of massive poverty reduction must be tempered somewhat by recognition of the limitations of the official definition of poverty used—a definition that was drawn up by the World Bank in the 1970s. The basis of the poverty measure is the adequacy of income for meeting minimal nutritional needs plus a specified amount for expenditure beyond this. Nutritional requirements change over time as the population changes (e.g. becomes taller), but in spite of this the NSO had not as of the mid-1990s changed the nutrition component of its poverty lines from the one originally used (Medhi 1996). Updating just this component of the poverty line calculation, Medhi Krongkaew finds that the poverty rate as of 1994 was 14.3 per cent (Kakwani and Medhi 1996; Dixon 1999: 218)—substantially higher than the official figure, though still a significant reduction from the 32 per cent recorded in 1975–6.

Several other points are worth making regarding these figures. First, though both the official poverty measure and Medhi's revised measure include a specified amount of income needed for non-food items, the method of determining this is problematic, simply using the average share of total income that households in the lowest quintile of the income distribution report spending on non-food items to determine how much households need for minimum non-food requirements. This is a fundamentally question-begging technique, since it makes no attempt to determine whether households in the bottom quintile *need* to spend more on non-food items than they actually are able to spend. The reason such a method is used is simply because no surveys which could establish minimum non-food needs have been undertaken (Medhi 1996: 24). This is a very non-trivial problem in a context of rapid economic change, where items that were once luxury items (e.g. motorcycles, telephones) can quickly become fairly standard requirements for full participation in social life (Sen 1987: 18). It is quite possible that the official poverty statistics are counting as non-poor a significant number of people who haven't been able to keep up with the expanded needs created by commodification.

Second, even if one sets aside these measurement concerns, the substantial poverty that remained even after three decades of rapid economic growth appears rather 'hard core' and intractable, suggesting that the process of economic growth in Thailand will not be adequate by itself to prevent millions of people from living lives of destitution, even in the midst of increased abundance (Medhi 1996: 8–9). Third, many of those who have entered the ranks of the non-poor over the recent decades of development have done so only marginally. Thus, economic slowdowns, increases in rice prices, and other such phenomena can quickly increase the number of people in poverty quite dramatically (Dixon 1999: 220).

head count measure. For example, if one measures the gap between incomes of the poor and what is required to reach the poverty line, Bangkok's problems seem far more severe than they do under the head count measure (Medhi 1996: 42).

5.3.2 *Economic growth and changes in the physical quality of life*

Having introduced these qualifications regarding the phenomenon of poverty reduction, it can nonetheless be meaningfully asserted that the material standards of living for most of the Thai population have improved over the post-World War II period. A useful way of further analysing this issue is to critically examine direct indicators of material betterment. Former USAID official Robert Muscat asserts 'by the usual measures of change in the material conditions of life, and in the health and educational status of the population, economic development has brought substantial improvement' (Muscat 1994: 238). Muscat's claim is defensible, but a careful examination of Thailand's performance in regional and international perspective brings the material improvements here into somewhat sharper focus.

Taking the aggregated data as a starting point, there can be little arguing with the assertion that basic social indicators have improved substantially: between 1960 and 1997, life expectancy at birth in Thailand increased from 52.3 years to 68.8 years, while the infant mortality rate per 1,000 live births decreased from 103 to 31. The mortality rate for children under age 5 decreased from 102 per 1,000 live births in 1970 to 38 in 1997. The adult literacy rate increased from 79 per cent in 1970 to 94.7 per cent in 1997 (UNDP 1993–9).

However, to infer from these figures, as do Muscat and others, that rapid economic growth provides the key to rapid improvement in human well-being is problematic (cf. Sen 1999: 43–6; UNDP 1999: 129). A more careful and comparative examination of the Thai data illustrates this point. For example, while Thailand's improvement in life expectancy between 1960 and 1997 was substantial, it was actually no more impressive than the increase for all countries categorized by the United Nations as 'medium human development' countries, almost all of which grew much more slowly than Thailand yet whose average longevity increased from 48.5 to 66.6 years (UNDP 1994: 136; 1999: 134).

The United Nations Development Program (UNDP) attempts to capture national performances in these social indicators with an index, which it calls the Human Development Index (HDI). The index is problematic in a variety of ways, but using some of its components to interrogate the Thai case provides a useful way of examining the claim that stellar Thai economic growth has produced stellar social performance.[4] The UNDP's 1994 report ranks Thailand as one of the star performers in human development, having registered the fifth largest increase in HDI value among all countries between 1960 and 1992, and the second largest between 1980 and 1992 (UNDP 1994: 96). The index includes performance in GDP, however, and if one disaggregates to examine performance in non-GDP components of the HDI—namely life expectancy at birth and

[4] The UNDP changed the way in which it constructs the HDI for its 1999 report. Thus, in the discussion that follows, I limit some comparisons to the comparable figures given in earlier reports. Where possible, however, I utilize the most recent data released in the 1999 report.

education—a different picture emerges. Thailand's economic performance is unquestionable: as noted earlier, it has been one of the 10 fastest growing economies in the world since the 1960s, and by 1997 it ranked 58th out of 174 nations in GDP per capita. However, if one ranks Thailand on the social components of the HDI, it falls to 88th out of all 174 nations compared in 1997.[5] In terms of improvements in the social indicators between 1960 and 1993, Thailand ranks 25 out of 84 developing nations—a creditable but scarcely miraculous achievement. Indeed, Thailand's performance appears similarly undramatic in several of the UNDP's more recently constructed poverty measures. In the Capability Poverty Measure, which is based on attendance of trained health personnel at births, percentage of underweight children under age 5, and female illiteracy rates, Thailand's rank for 1993 was 30th out of 101 developing nations (UNDP 1996: 111). In the Human Poverty Index—which is based on the percentage of people not expected to survive to age 40, the adult illiteracy rate, the percentage of the population without access to safe water, health services, and sanitation, and the percentage of underweight children under age 5—Thailand's rank for 1997 was 29th out of 92 developing nations (UNDP 1999: 146–8).

5.3.3 Economic growth and growth in economic disparity

I have argued that the specific pattern of rapid economic growth in Thailand has been predicated to a significant extent on the creation and exacerbation of enormous social and economic disparities. These, in turn, have much to do with the failure of the Thai political economy to translate rapid economic growth into commensurate, broad-based social gains. This is particularly evident in the area of education, which has received less emphasis from the Thai state than warranted, especially in rural areas (Bello, Cunningham, and Poh 1998: 56–7; Dixon 1999: 92). It is also evident in the performance on human health indicators, which though respectable in the aggregate is badly wanting in areas such as AIDS and accident prevention, which disproportionately affect the poor and the working class (Bello, Cunningham, and Poh 1998: 83, 120; Glassman 2001).

Disparities that limit improvements in the education and health of the less privileged reflect the economic marginalization discussed in previous chapters. This is particularly the case because the Thai economy has become more highly monetized while state-funded social service provision has remained underdeveloped (Nikom 1995: 10–11). In a cash economy where little social service is provided, relatively substantial cash incomes may be necessary for procuring access to educational, health care, and other services beyond the bare minimum necessary for subsistence. Thus, unravelling the reasons for less than spectacular social performance in one of the world's fastest growing economies involves examining the relatively marginal economic position in

[5] Thailand's ranking in the Physical Quality of Life Index (PQLI) for 1995 was 70 out of 178 (Porter and Sheppard 1998: 9–13).

which development has left much of the population, in spite of the aggregate economic and material improvements noted above. Here, I examine economic marginality through the use of data on income distribution.

Several broad indicators can be used to demonstrate the evolution of income disparity in Thailand from 1963 to 2000. The Gini index indicates disparities among all households, with a bias towards emphasizing changes in the middle of the income distribution (Atkinson 1983: 53–9). The ratio of income between the top 20 per cent of income earners and the bottom 20 per cent provides a useful counterpoint to the Gini index by emphasizing changes at the extremes of the income distribution. By these measures, income distribution in Thailand has become noticeably more inequitable over the 37 years in question, as seen in Table 5.1. Moreover, the gap expanded particularly quickly after 1981, when the average income of the top decile group tripled, as they garnered one-half of all income gains (Pasuk and Baker 1998: 285).

While absolute numbers on income distribution are dangerous to compare across countries—given differences in collection methods and data quality—it is important to attempt to place Thailand's disparities in some kind of comparative perspective. According to evidence summarized by Voravidh Charoenloet (1996), the ratio of the richest 40 per cent of Thai income earners to the poorest 60 per cent made the country roughly the sixth most inequitable in the world by the late 1980s and early 1990s. While none of these income distribution estimates is by itself decisive, collectively they provide evidence for the claim that rapid economic growth in Thailand has led to an erosion of economic equality over time and has made Thailand one of the more—not one of the less—inequitable countries in the world.[6]

Table 5.1. Income distribution in Thailand, 1963–2000

Year	Gini coefficient	Top 20/Bottom 20[a]
1963	.414	6.3
1975	.426	8.3
1981	.453	9.5
1988	.485	11.8
1990	.522	13.7
1992	.536	15.0
1994	.497	14.4
1996	.516	13.5
2000	.525	14.8

[a] Income of the richest 20 per cent divided by income of the poorest 20 per cent.

Sources: NSO, *Household Socio-Economic Survey* (1975–94); Medhi (1996); Kakwani and Medhi (1996); Mingsarn (2003)

[6] The NSO's official data on income distribution, reported in the *SES*, imply a different picture—one in which economic inequality has been declining steadily since the mid-1980s. These data, however, are inaccurate, using population weights that do not correspond to the actual population of the country. Medhi Krongkaew and Nanak Kakwani have thus developed revised population weights which yield the Gini coefficients and 20/20 ratios recorded here for 1988–2000 (Medhi 1996: 7).

These data can now be broken down in several ways to give a clearer picture of sub-national disparities that have emerged or been exacerbated in Thailand through the process of economic growth. Table 4.6 showed changes in GRP over the years 1978–96 and illustrated the economic weight the BMR and the Central Region have taken on in the national economy. In this obvious spatial sense, rapid economic growth has been somewhat less a Thai phenomenon than a Bangkok-centric phenomenon, even though growth has clearly occurred elsewhere and has transformed local social structures.

The class and gender relations that articulate these spatial disparities are particularly important to sketch, even though the existing primary data do not allow us to do this as well as would be desirable. Class-based data on income distribution, in the Marxist sense, are not available; but we can nonetheless infer some things about the class basis of economic growth in Thailand by examining the distribution of income across occupational categories.

Table 5.2 shows *SES* data on changes in incomes for various occupations over the years 1986 to 1996. Several features of this table are noteworthy. First, there is a large and obvious skew of income between people in the BMR and the rest of the country by 1996. Second, while disparities between Bangkok and the rest of the country are important, disparities are also growing between, on the one hand, owners of capital and urban professionals (represented by the categories 'entrepreneurs' and 'professionals') and, on the other hand, peasants and industrial workers (represented by the categories 'farm operator, mainly renting' and 'production and construction workers'). Table 5.3—which shows how the incomes of all groups compare to the national average—indicates that the spatial location of labour strongly affects how its relative position in the national economy evolves, but it also indicates that even *within* the BMR, sanitary districts, and villages, workers have seen their relative positions decline. The only exception to this is production and construction workers in municipal areas outside of Bangkok, whose position relative to other municipal employees improved slightly, despite a relative decline in national terms. It appears, then, that the growth process has strengthened the overall position of capital and management relative to labour and the peasantry, though it has actually weakened the position of most upcountry entrepreneurs and professionals relative to those in Bangkok. It has also undermined slightly the position of the Bangkok proletariat while strengthening the local (but not national) position of urban workers elsewhere.

The gender dimensions of the income disparity evidence are even somewhat more difficult to elucidate with the existing data than are the class dimensions. *SES* data indicate no important differences in average household income according to whether the head of household is male or female (Suganya and Somchai 1988: 27; Medhi 1996: 30), but in a sense this is a fairly unsurprising finding in a society where inheritance is usually matrilineal. We cannot hope to discern much about disparities between men and women without disaggregating the household data. One way of doing this is simply to look at the *LFS* data

Table 5.2. Per capita income by occupation and community type, 1986, 1996 (baht per month, constant 1988 prices)

Occupation	All areas	BMR[a]	Municipal Areas	Sanitary districts	Villages
1986					
All households	928	1,941	1,955	1,087	662
Farm operators, mainly owning land, over 40 rai	813	1,391	3,714	892	811
Farm operators, mainly renting land, 5–19 rai	448	769	1,035	502	436
Entrepreneurs, with paid employees	2,747	3,761	3,696	2,420	1,581
Professional workers, own account	1,685	1,197	2,026	3,688	1,183
Professional workers, employed by others	2,565	3,552	3,015	2,512	1,987
Farm workers	468	771	1,529	507	421
General workers	535	1,328	796	579	464
Production and construction workers	1,103	1,506	1,157	993	859
1996					
All households	1,945	4,620	3,329	2,055	1,302
Farm operators, mainly owning land, over 40 rai	1,788	2,833	8,141	4,670	1,611
Farm operators, mainly renting land, 5–19 rai	791	1,163	1,275	927	774
Entrepreneurs, with paid employees	5,596	8,981	6,013	4,781	3,092
Professional workers, own account	5,092	5,821	4,621	4,067	1,930
Professional workers, employed by others	5,254	8,457	5,345	4,433	3,211
Farm workers	835	1,298	1,581	1,041	798
General workers	1,019	2,212	1,448	1,009	881
Production and construction workers	1,520	3,009	2,030	1,368	1,156

[a] *Household Socio-Economic Survey* data calculated only for Bangkok, Nonthaburi, Samut Prakan, and Pathum Thani.

Source: Calculated from NSO, *Household Socio-Economic Survey* (1986; 1996).

Table 5.3. Relative per capita income by occupation and community type, 1986, 1996 (income as % of national average, 1988 prices)

Occupation	All areas	BMR[a]	Municipal Areas	Sanitary districts	Villages
1986					
All households	100	216	209	116	71
Farm operators, mainly owning land, over 40 rai	88	155	398	95	87
Farm operators, mainly renting land, 5–19 rai	48	86	111	54	47
Entrepreneurs, with paid employees	296	419	396	259	169
Professional workers, own account	182	133	217	395	127
Professional workers, employed by others	276	395	323	269	213
Farm workers	50	86	164	54	45
General workers	58	148	85	62	50
Production and construction workers	119	168	124	106	92
1996					
All households	100	238	171	106	67
Farm operators, mainly owning land, over 40 rai	92	146	418	240	83
Farm operators, mainly renting land, 5–19 rai	41	60	66	48	40
Entrepreneurs, with paid employees	288	463	309	246	159
Professional workers, own account	262	300	237	209	99
Professional workers, employed by others	270	436	275	228	165
Farm workers	43	67	81	53	41
General workers	52	114	74	52	45
Production and construction workers	78	155	104	70	59

[a] *Household Socio-Economic Survey* data calculated only for Bangkok, Nonthaburi, Samut Prakan, and Pathum Thani.

Source: Calculated from NSO, *Household Socio-Economic Survey* (1986; 1996).

on incomes for men and women in different economic sectors and in different regions. Overall, income gaps between men and women have closed a little during the years 1983 to 1997. In municipal areas, women's wages as a share of men's increased from 71 to 78 per cent in manufacturing, 68 to 87 per cent in commerce, 59 to 65 per cent in services, and 67 to 69 per cent in construction. In municipal areas of the North, where the change was the most dramatic, women's manufacturing wages as a share of men's increased from 66 to 80 per cent (NSO, *LFS*, February Round 1983–97). Nonetheless, as these numbers indicate, gender-based income disparities remained significant by the end of the boom.

Other considerations have to be entered into the picture as well. The full implications of increased incomes for women as a group cannot be discerned unless we know to what extent these incomes indicate increases associated with women entering the paid labour force for the first time and to what extent they indicate increases in wages for women already employed. The very rapid rise in wages for women in rural manufacturing industries of Northern Thailand—from 55 to 69 per cent of men's wages (NSO, *LFS*, February Round 1983–97)—gives some indication that first-time entry into the wage labour force may be an important contributor to statistical increases at the aggregate level.

This in turn gives rise to other questions about the transformation of women's labour in the process of industrialization. Perhaps the most basic for determining the equity effects of new income streams has to do with how many hours women are working in unpaid household labour as they enter the paid labour force. If the hours of labour women begin putting in at the factory are compensated by an equal reduction in hours of household labour then, at least in the narrow terms of national accounting, the increased income streams are a straightforward gain in economic well-being. But to the degree that women retain already inequitably distributed responsibilities for household labour as they enter the manufacturing workforce the story is less clear-cut. If women's new streams of income are all given over to the household as a whole while women continue to perform household chores in the same manner as before, then the net equity effects for women of the new income streams might in fact be negative.

Unfortunately, it is impossible to give much quantitative substance to these theoretical considerations, since there have not been an adequate number of empirical studies done regarding the effects of manufacturing employment on women's work hours, household responsibilities, or control over new income streams. Nonetheless, by pulling together some sketchy evidence, we can illustrate possibilities. Using data from Varunee and Benja's study, it can be shown that if most of a woman's added income from working outside the home goes back to the household as a whole—as is overwhelmingly the case for women working in food-processing factories and largely the case for women working in textile and electronic factories (Varunee and Benja 1994: 66)—and is divided equally among all household members while household labour responsibilities remain predominantly those of women (p. 137), then women's wages per hour for all labour undertaken increase only slightly from working outside the home while men's wages increase more substantially merely as a result of women's added household contributions. For example, one set of calculations makes the following assumptions: women employed 48 hours a week in food-processing factories work 35 additional hours per week in the home while women not employed outside the home work 45 hours per week in the home; men whose wives work in food-processing factories contribute 20 hours per week to household chores while men whose wives don't work in factories contribute 10 hours per week to such chores; women in food-processing make a monthly income of 3,036 baht (p. 134) while their husbands make 2,900 baht per month, working 48 hours a week (p. 135); household expenditures are for the most part decided upon jointly (pp. 138–9) and money and consumption goods are divided evenly among household members. Under these assumptions, women who take on jobs in food-processing while performing the 'double shift' see their effective hourly wage for all work done rise from approximately 7.7 baht per hour to 8.5 baht per hour while their additional 38 hour-per-week labour contribution helps raise their husband's effective hourly wage for all work done from 6.0 baht per hour to 10.4 baht per hour.[7]

On the basis of this sort of conjectural evidence and the qualitative discussion presented in Chapter 4, what can be suggested are several general points. As women enter fields such as manufacturing labour for the first time the effects of this on their lives vary somewhat according to factors such as their age, marital status, whether or not they have children, whether they are from urban or rural areas, and whether they migrate to find work or remain close to home. It appears that in narrow income terms, the effects of manufacturing employment on younger, better-educated, and unmarried women who can find jobs in the higher-paying factories close to urban centres may be more favourable than are the effects on older, less-educated women with children who have more household responsibilities and are less able to either relocate to areas with higher paying employment or spend long periods of time away from home (Gray 1990; Varunee and Benja 1994).

In sum, we can say that the income equity effects of rapid economic growth for women have been ambiguous. Some women have clearly gained from the process—particularly if they have been part of the BMR's pool of entrepreneurs or urban professionals—but for most other women the precise effects depend on the specificities indicated above. Despite these ambiguities, however, the fact that class disparities have grown substantially in Thailand since

[7] The assumptions about number of hours worked are based on figures cited by Bell (1997: 70), which show that women in Asia spend between 36 and 45 hours a week in unpaid housework, and do 80–90% of all the labour. I assume that women working outside the home fall toward the lower end of the range and women not working outside the home toward the upper end. I make a slightly more generous assumption about the number of hours worked by men in the home.

at least the 1960s is quite important for the gender implications of growth, since manufacturing employment itself is highly gendered. Given that women supply most of the labour in the leading export industries that have triggered rapid economic growth, and given that women working in the sex trades have also contributed substantially to foreign currency earnings (Pasuk, Sungsidh, and Nualnoi 1998), growth with equity ought to imply that women would make proportionately larger income gains for their highly productive labours. This clearly has not happened, and in this sense one can argue that in spite of real income growth for women in the aggregate, such growth has not improved equity but has in fact been consistent with increasing exploitation of women as labourers (Bell 1992; 1996; 1997).

5.4 Reconsidering the East Asian boom: 'miracle', managed growth, or uneven internationalization?

Neo-Weberian analyses of East Asian economic growth have seriously undermined the credibility of earlier neo-liberal accounts that attempted to explain such growth in terms of an export oriented industrialization project overseen by a putatively minimalist state. Forced to concede that states such as those of South Korea and Taiwan did in fact 'intervene' strongly to promote industrial transformation, neo-liberals then shifted their arguments to the strained and problematic assertion that all of the crucial state 'interventions' pointed out by neo-Weberian scholars have been 'market conforming' (World Bank 1993; cf. Wade 1996*b*).

Yet if neo-Weberians have effectively challenged neo-liberal canons, neither they nor neo-liberal scholars have provided an adequate framework for understanding many of the aspects of industrial development discussed in the previous chapters. While the developmental state paradigm—particularly as presented by Johnson in his analysis of the Japanese state—has proven very useful for identifying possible and relatively successful forms of state industrial policy that are at odds with the neo-liberal 'Washington consensus', this same paradigm has proved to be a straitjacket when applied rigidly as a formula for explaining region-wide East Asian growth. Thailand and, for the most part, other Southeast Asian NICs, do not fit comfortably into the developmental state mould (Jomo 1997), and even the South Korean state may not fit this mould as fully as originally advertised in the neo-Weberian literature (Chibber 1999). What has become clear, therefore, is that the East Asian growth dynamic is not guided by any singular type of state policy but rather involves shared experiences of growth across states with substantially distinctive forms, functions, and policy orientations. From this it follows that the policies of developmental states—whether those of Japan, or the slightly more constrained developmentalism of the Northeast Asian NICs—cannot provide the key that unlocks the riddle of shared East Asian growth. Instead, as suggested here, it appears that shared East Asian experiences of rapid growth and industrial transformation must be explained largely in terms of processes that link producers in the region to one another, across their local differences, as part of a regionalization pattern that is integral to broader patterns of internationalization (Bernard and Ravenhill 1995).

Both neo-liberal and neo-Weberian analyses run headlong into the limits of attempts to explain such regionalization and internationalization processes from within the 'territorial trap', which prevents either approach from adequately conceptualizing how states are continuously transformed in relation to each other and to internationalized class processes. This same weakness also makes it difficult for neo-liberals or neo-Weberians to adequately address the sub-national socio-spatial unevenness of capitalist growth and industrial transformation. Disciplining of labour, for example, features only marginally in discussions by either neo-liberals or neo-Weberians. The World Bank's *East Asian Miracle* report notes in passing that East Asian NICs repressed labour unions (World Bank 1993: 164–6, 270–2), but this is rationalized by the report's assertion that labour policies prevented unions from 'distorting' wages upward (World Bank 1993: 266). The report nowhere attempts to elucidate the social disparities to which such labour practices have led—which it incorrectly claims are minimal—and instead assumes labour repression to be part of a process producing broadly shared national benefits.

Neo-Weberians have been somewhat less circumspect in recognizing labour repression, but such recognition as has been accorded to this phenomenon has played little role in their analyses of economic growth or industrial transformation. Instead, neo-Weberians have focused primarily on the state's attempts to discipline capital, making this the cornerstone of the developmental state paradigm. Indeed, in accounts of the developmental state such as that of Leftwich (1995), a 'weak' and subordinated civil society is assumed at the outset as a condition of developmental state success rather than being shown to be a (tenuous) *product* of concerted state efforts to repress sometimes unruly peasant and labour organizations. Here, as in neo-liberal analyses, the internal social spaces of the national territory are portrayed as more homogeneous, unified, or uncontested than they have been in reality.

Such 'territorially trapped' conceptualizations do not merely pose problems for analyses of income distribution and other indicators of social equity and uneven development. The process of East Asian industrial transformation can itself be better understood from outside the territorial trap. Not only have regionalized production networks and state alliances helped stimulate shared processes of growth and economic transformation, but class struggles and uneven development both within and across national boundaries have helped generate these regionalized production networks, as when Northeast Asian capitalists move to Southeast Asia to fight rising wages at home, and when Southeast Asian workers migrate to higher wage countries in the region and remit wages. Even the NIC states that best fit the developmental state mould—such as South Korea (Cumings 1984) and Singapore (Rodan 1989)—were not formed through narrowly national projects but rather in relation to regionalized class (-relevant) struggles connecting (in complex ways) rebellious Chinese, Korean, and Vietnamese peasants, communist parties, regional economic and political elites, and capitalist Cold War allies. The industrial transformations that unevenly spread across East Asia after World War II—from Japan to the Northeast Asian NICs and eventually to Southeast Asia—were part and parcel of this complex, internationalized class (-relevant) struggle, which took place both within and across states, shaping each of their nationally and sub-nationally specific structures. In this sense, the state actors whose activities fomented industrial transformation in East Asia were neither laissez-faire friends of an abstract 'free market' nor plan-rational friends of an abstract 'governed market' but rather partisans in social struggles. It was these struggles that unevenly produced internationalization of capital and industrial transformation, through the place-specific but interconnected class transformations that occurred within and across each of the East Asian states.

5.5 Conclusion

The rapid economic growth that prior to 1996 called attention to a Thai development 'miracle' was palpably international and palpably fragmentary. These two aspects of economic growth in Thailand are in fact interrelated. The ways in which both capital accumulation and state power in Thailand are linked to processes of internationalization help explain, on the one hand, the participation of the Thai political economy in a broader East Asian growth phenomenon, and, on the other hand, specific features of Thailand's internally uneven and inequitable development.

The strength of Thailand's aggregate national economic growth performance relative to that of developing countries in other parts of the world reflects geographical-historical realities that neither capitalists nor states in any particular locale called into being. Rather, in a long-term process of geographical expansion, capitalists and capitalist states both in East Asia and outside of it created political economic realities that raised considerably the potential for rapid growth within the region. States were scarcely passive bystanders in this process since the ways in which they were transformed and linked into international state configurations determined much about their access to internationalizing streams of capital—including those flowing from international lenders or national aid agencies—and international markets. The state in Thailand was successfully transformed during the post-World War II period into a firmly anti-communist state that facilitated capitalist expansion and made industrial labour available to capital on acceptable terms. As such, it was positioned both to help 'domestic' capital gain access to international markets and to receive substantial flows of FDI, which helped trigger more rapid industrial growth within Thailand. Such positioning by the state did not necessarily involve all of the more sophisticated strategies for developing indigenous manufacturing capacity favoured by partisans of the developmental state, but neither did it simply involve sitting back and letting markets do the work, since neither the creation of basic conditions for market growth nor the generation of a suitable industrial proletariat are tasks performed by market forces themselves. The state in Thailand performed these tasks ably enough to stimulate rapid industrial growth and transformation, but it did not do so as a narrowly national entity; rather, one might say that internationalization of capital in Thailand helped generate and was generated by an internationalization of state forms and functions—again, as an integral part of broader regional patterns evident throughout East Asia.

These processes of internationalization in the post-World War II period interacted with and transformed various pre-existing socio-spatial structures to give development in Thailand its particular character. Well-developed class structures and geographical concentrations of power and wealth were not undermined but were rather enhanced in transformative ways. More fully-fledged industrial growth increased Bangkok's sub-imperial domination of outlying regions—albeit not without a significant effort by the state to undermine rural rebellion—and also increased the relative wealth and power of elites who had transformed themselves into the rulers of a modern capitalist state. Not only did the techniques of modernization and governance these elites adopted as they transformed themselves owe much to their international alliances, but their very ability to maintain adequate political stability and an environment conducive to economic growth while disparities were being aggravated owed much to their international support. Both state capacity and state autonomy, in other words, were less the products of insulation and determination of elite bureaucrats—as generally insisted by neo-Weberians (Leftwich 1995)—than they were the products of the internationalization of the state and the specific capacity-building projects undertaken by various transnational coalitions as part of this process.

The Thai case, seen in this way, suggests a variety of problems with the prevalent ways of measuring economic performance in the global economy, which tend to ignore both the interrelated character of development processes across national borders and the unevenness of development at sub-national scales. Mere comparative statistics at the national level do not adequately reveal how economic growth processes within different nationally conceived social formations might actually be connected with one another. The important connections must be supplied by a geographical-historical analysis that draws out the processes by which capital accumulates and by which state forces act in different locations. Moreover, national statistics mask internal differentiation and thus differences in the meaning of aggregate GDP growth for human populations in different rapidly growing economies. In short, conventional, territorially trapped methods of national comparison fail to recognize the ways in which development is both more and less than a national phenomenon.

In conclusion, it can be noted that to highlight the role of internationalization processes in Thailand's growth is not to provide a master key for interpretation of all development phenomena. Escaping the territorial trap simply enables one to view more clearly varied factors contributing to the phenomena in question. For example, internationalization of the state and the economy is not by itself a process that differentiates Thailand from other countries in East Asia, all of which have undergone forms of internationalization. Indeed, as argued here, it is the *interconnection* of these East Asian regional processes of internationalization that has perhaps contributed the most to the region's broad success in the economic growth sweepstakes.

Yet internationalization occurred in different ways within different places, and attention to the different rhythms of internationalization in these different places helps explain different outcomes. For example, the Northeast Asian NICs began their post-World War II export growth drives somewhat earlier than the Southeast Asian NICs, and were constrained by natural resource limitations to focus those drives from the beginning on the development of manufactured goods. The Southeast Asian NICs, by contrast, could rely for a longer period of time on natural resource exports and thus began developing manufactured exports long after the Northeast Asian NICs had advanced up the industrial ladder (Jomo 1997)—a matter of great significance for the Southeast Asian NICs since it has meant that they face a greater array of competitors in their bid to industrialize on the basis of export growth.

Moreover, the internationalization of the state in East Asia has occurred in a variety of ways that are related to these variations within the regional pattern of economic growth. For example, post-World War II US involvement with South Korean and Taiwanese leaders occurred after substantial peasant rebellions were already at hand in Korea and Vietnam, and with the Chinese revolution fully in motion. This meant that the position of the Cold War allies in East Asia was relatively precarious and led to willingness on their part—abetted by specific geographical-historical conditions affecting the local class structure—to promote land reform as a co-optive measure (Cumings 1984). In Thailand, by contrast, the US-Thai Cold War alliance was cemented long before peasant rebellions became an important force within the country, and by the time such rebellions emerged the alliance was strong, heavily militarized, and deeply opposed to any co-optive strategies for peasants. Thus, land reform has never been on the agenda in post-World War II Thailand (Bello, Cunningham, and Poh 1998; Dixon 1999).

In short, international forces that come more firmly into view outside of the territorial trap do not by themselves fully explain growth patterns or political processes in Thailand or elsewhere in East Asia, these patterns always being mediated by a variety of local forces and processes. But what highlighting internationalization of capital and the state in East Asian growth does is both to help explain what sets the region as a whole apart from other places in the world in its growth dynamism and to help explain the specific international connections that mediate differences between the patterns of growth and political struggle within various countries of the region. Moreover, as will now be argued, these international connections strongly mediate the locally varied processes of economic downturn that began in the late 1990s.

6

Uneven Economic Crisis, Industrial Restructuring, and the Politics of Development in a Post-Nationalist Era

6.1 Introduction

Up until 1996, the interpretation of Thailand as something at least akin to an economic 'miracle' was hegemonic, and the analysis offered here has not so much attempted to challenge this interpretation on its own terms as to offer an alternative interpretation that recognizes why Thailand's rapid industrialization process can be considered simultaneously a success in capitalist terms and a highly troubled process from other perspectives. The economic events that began in 1996, however, have considerably tarnished the mainstream image of Thailand as a 'new little dragon' and have called into question the notion that Thailand's development has been an unquestionable *capitalist* success story. The framework I have presented for analysing Thailand's uneven and contradictory success story can also be used to analyse the dynamics that are at work in the recent bout of economic bust, partial recovery, and post-crisis political manœuvring.

In this chapter, then, I round out the discussion of Thailand's uneven industrial transformation and the role of the Thai state in this process by suggesting how the dialectically conceived internationalization processes discussed earlier might help to explain the nature of the contemporary economic crisis and the economic challenges that lie ahead. More specifically, I offer here an alternative to the dominant explanations of the Thai economic crisis, which have tended to focus narrowly either on corruption and lack of transparency in the functioning of Thai institutions (the dominant line of analysis emanating from the West and neo-liberals) or on international forces beyond the control of the Thai state, such as currency traders and IMF measures (a prominent line of analysis in much of Southeast Asia and among neo-Weberians). While these lines of analysis vary in where they place the blame for the Thai crisis, they share the view that the crisis is primarily financial and does not reflect deep, underlying structural problems in either the Thai pattern of industrial growth or the place of small industrial exporting countries in the global economy.

The analysis I offer here differs from both types of views on a number of counts.[1] First, it suggests that neither 'internal' forces (the Thai state, Thai capital) nor 'external' forces (international capital, the IMF) can be rightly seen as independent sources of economic malaise. Rather, as partners in the internationalization of capital and of the Thai state, they have interactively shaped the Thai crisis. Second, this crisis is not merely a financial crisis but reflects serious limitations *of capital itself* in the favoured model of development in Thailand. Thus, it is argued, the Thai financial collapse of 1996–7 was not a thunderbolt out of the blue but rather the maturation of deeply rooted crisis tendencies integral to the capital accumulation process—tendencies that were already becoming visible by the early 1990s, even before the financial press noted the serious warning signals of slumping real estate markets, declining stock prices, and pressure on the exchange rate.

Moreover, it is argued here that the very features of Thai development policy that had been praised by the World Bank and IMF and which had facilitated rapid growth (repression of wages, reliance on exports and foreign direct investment, relatively limited state discipline of capital) have facilitated economic crisis in the changing global environment of the post-Cold War era. Thai development manifests the unevenness and contradictory character of capitalist industrial growth, including the propensity towards cyclical downturn; and Thailand's internationalist orientation links the working out of this 'internal' propensity directly to the uneven internationalization of capital.

Moreover, as I highlight in the second part of this chapter, the Thai state's deep embeddedness in global neo-liberalism creates contradictions in the project of rebuilding the economy. Immediately following the crash, the Democrat Party seized the political and economic helm, with strong backing from the international financial community and internationalized capital more generally, implementing the IMF's structural adjustment programme (SAP) for Thailand with considerable commitment. The SAP did not lead to a robust or rapid economic recovery, and by the end of 2000, many Thais had wearied of neo-liberal policies and voted for a new party, Thai Rak Thai, which promised more government stimulus programmes to spur recovery. Thai Rak Thai has been characterized by some international observers as a more nationalist and populist party than its predecessors yet, as will be shown, the rise of Thai Rak Thai is a more complex phenomenon than the mere expression of nationalist backlash, and indeed many of the party's leading members and supporters are such integral members of Thailand's internationalized capitalist elite that Thai Rak Thai makes a very poor candidate for a nationalist party. As such, Thailand has now seemingly passed so far down the internationalist road that it is part of a fundamentally 'post-nationalist' global political economy, and any form of nationalism it undertakes to promote is thus likely to be limited at best.

[1] For an analysis that parallels some of the arguments here, see Bernard (1999). See also, Glassman (2003); Glassman and Carmody (2001).

6.2 Economic meltdown in Thailand

In illustrating the relevance of these theoretical considerations, I analyse the economic crisis in Thailand as a simultaneously national and international phenomenon. In doing this, I show how the national and international dimensions of the crisis were both integrally interrelated and in fact mutually constitutive of one another. I thus highlight the futility of narrowly ascribing responsibility for the crisis to either 'internal' or 'external' causes. Throughout the discussion, I pay much attention to the role of the state in the Thai crisis, while highlighting the importance of class and class-relevant struggles over the distribution of surplus.

6.2.1 The outbreak of economic crisis in Thailand

A central theme in Marxist theories of economic crisis is the movement of the profit rate, analysis of which involves a variety of complications, both theoretical and practical. In the former category is the issue of whether or not to measure the profit rate in price terms (Devine 1994) or in value terms (Shaikh and Tonak 1994). While the theoretical issues underlying this debate are significant, it turns out that empirical estimates of prices and labour values generally find that they are highly correlated (Sheppard and Barnes 1990: 50), so for analysis of the *trend* in the profit rate (as opposed to the absolute value), the choice of measures may not be significant. Profit rates are reported here in price terms, though the value rate of profit shows the same trend.

Conceptually, there are different ways of discussing profitability. Gerard Duménil and Dominique Lévy (1993) distinguish between the profit rate, the profit margin, and the profit share. The profit margin is the ratio of profits to production costs. As Duménil and Lévy note, this measure is useful in that it relates profits to the price of inputs such as raw materials, energy, and labour, but it does not measure the ability of a stock of capital to yield profit (Duménil and Lévy 1993: 21). Profit share refers to the percentage of the total value added in production that goes to capital (the remainder being the share paid out to labour in wages). This is a straightforward measure of distribution of the surplus, but it does not reflect the ability of capital to make a profit (Duménil and Lévy 1993: 20–1).

The profit rate refers to the relationship between profits and the capital stock—i.e. the amount of money invested in a particular line (Duménil and Lévy 1993: 20–2). Mathematically, it can be rendered by the formula

$$r = P/K = (Y - W)/K \qquad (6.1)$$

where r stands for the profit rate, P stands for profits, K stands for capital stock, Y stands for output (or value added), and W stands for total compensation (wages).

Figure 6.1 shows the movement of the profit rate in Thailand over the period from 1970 to 1996. The profit rate in manufacturing has shown an uneven longer-term trend, but rises slightly at the end of the 1980s and declines in marked fashion in the 1990s.[2] Notably, the profit rate in most sectors has declined along with those in manufacturing and construction, with one exception being the rates in banking, insurance, and real estate (BIR, aggregated into one sector in the National Accounts data), an exception that will be discussed below.

I do not infer here a long-term crisis of declining profitability rooted in increasing organic composition of capital—the most classically Marxist explanation for declining profit rates—since the available data do not seem to unequivocally support such an interpretation. Short-term changes in the manufacturing profit rate between the mid-1980s and mid-1990s, on the other hand, might be seen as corresponding to a business cycle, a view that is supported by the figures on capacity utilization in manufacturing, trade, and construction.[3] Capacity utilization in manufacturing increased from 74 per

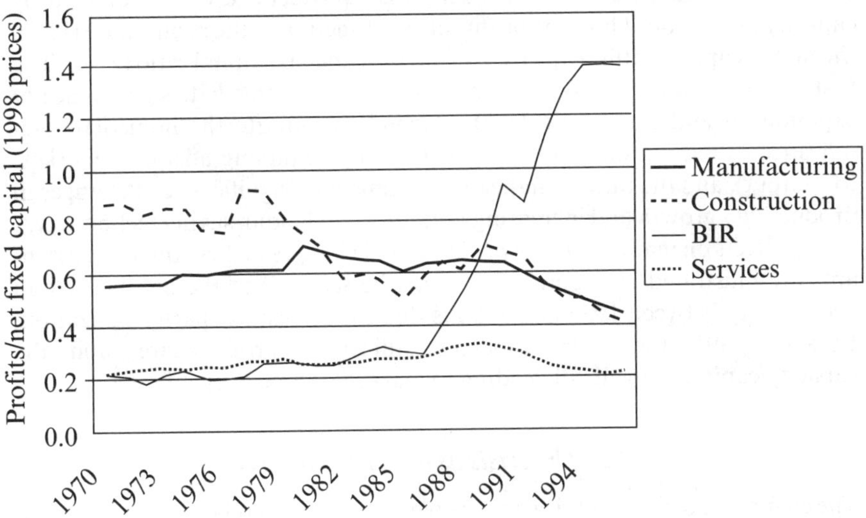

Fig. 6.1. Profit rates, 1970–1996 (profit rates = value-added minus employee compensation, divided by net fixed capital (at 1988 prices)).

Sources: Calculated from NESDB, *National Accounts* (1970–96); *Capital Stock of Thailand* (1970–96).

[2] A World Bank study of return on assets for all businesses in Thailand finds a similar pattern for the years 1988–96 (Claessens, Djankov, and Lang 1998).

[3] There is typically a strong statistical relationship between the profit rate and capacity utilization (Glyn 1997: 597), and the latter changes more in response to the business cycle than in response to longer-term trends.

cent in 1985 to 92 per cent in 1990 but then began a marked decline, dropping to 69 per cent by 1997 (Bank of Thailand, *Production, Investment, and Employment in Manufacturing, Trade and Construction Sectors* 1985–97).

The profit rate as defined by Duménil and Lévy can be decomposed into two parts, the profit share and the output–capital ratio:

$$r = (P/Y) \times (Y/K) \qquad (6.2)$$

where P/Y is the profit share and Y/K is the output–capital ratio. Profit rates will tend to decline if there is either a decline in the profit share (i.e. labour claims a larger share of the surplus) or a decline in the output–capital ratio (i.e. the output per amount of fixed capital stock decreases). Thomas Weisskopf (1979: 342) provides a method for further decomposing the profit rate that adjusts the output–capital ratio for changes in capacity utilization, an approach that can be used with the Thai profit series from 1977 to 1996. Weisskopf defines the profit rate as follows:

$$r = P/K = (P/Y) \times (Y/Z) \times (Z/K) \qquad (6.3)$$

where Z, potential output (or capacity), is equivalent to output divided by capacity utilization. Thus, the profit rate is a function of the profit share (P/Y), the rate of capacity utilization (Y/Z), and the capacity/capital ratio (Z/K). The first of these is taken by Weisskopf to indicate the immediate significance of exploitation and class struggle, the second to indicate the immediate significance of realization problems (or difficulty in finding adequate markets for output), and the third to indicate the immediate significance of changes in productivity growth and increasing investment in labour-saving technology.[4]

The effects on manufacturing profit rates of changes in these three determinants are illustrated in Figure 6.2. As can be seen, all of the net increase in profitability between 1985 and 1990 is due to increased capacity utilization. Between 1991 and 1996, capacity utilization, profit share, and the capacity/capital ratio all fall, leading to marked decline in profit rates.

6.2.2 *The crisis as a realization crisis*

The changes in capacity utilization shown here and their relationship to declining profitability indicate downward pressure on profit rates because of realization failure—the inability to find adequate markets in which produced commodities can be sold at prices that cover production costs and expected profit margins (Weisskopf 1979: 346). This interpretation is supported by evidence regarding developments impinging on the Thai economy from the late 1980s onward—particularly increased international export competition,

[4] Weisskopf acknowledges that all of these factors are interrelated and not strictly separable as causes of declining profitability. The point of disaggregation is thus only to sharpen discussion of the relationship between each of the various factors. On this issue see, also, Bell (1977).

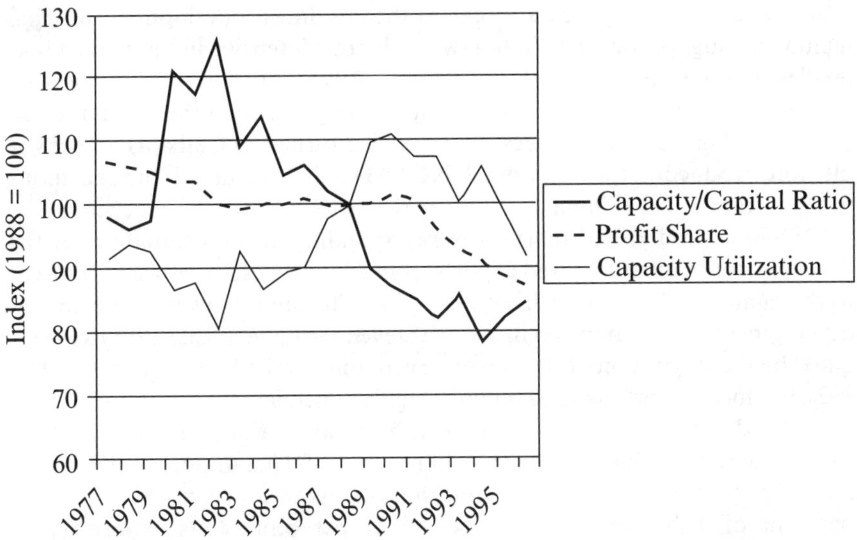

Fig. 6.2. Determinants of manufacturing profit rate, 1977–1996 (capacity/capital ratio = capacity divided by net fixed capital; profit share = value-added minus employee compensation, divided by value-added; capacity utilization taken from Bank of Thailand estimates).

Sources: Calculated from NESDB, *National Accounts* (1970–96); *Capital Stock of Thailand* (1970–96); Bank of Thailand, *Production, Investment, and Employment in Manufacturing, Trade and Construction Sectors* (1985–97).

which has made it more difficult for Thai and other Southeast Asian exporters to expand their share of foreign markets, even as their output (and that of competitors) increased dramatically. Thailand and its East Asian export competitors have collectively participated in creating this increased export competition, with each of them engaged in relatively disarticulated and foreign-market dependent growth. This growth, moreover, has involved relatively limited complementarity between the Asian exporters, thus intensifying the competition for markets outside of the region. As a consequence, there has been a burgeoning contradiction between regional overcapacity and tightening global markets, which places increasing pressure on all of the Asian export economies (Brenner 1998; Bernard 1999; Glassman 2003).

The lower-end Asian exporters have been particularly profoundly affected by one of the major economic events of the late twentieth century, the rise of China as an export manufacturing power. The importance of overseas Chinese networks of capital has been discussed in much recent work on Asian economic growth (e.g. Hsing 1998; Dicken and Yeung 1999). While in the past this network has been seen as facilitating economic growth in Thailand (Pasuk and Baker 1997), the opening of China to capitalism and the movement of large amounts of investment by Hong Kongese and Taiwanese capitalists into southern China now poses a threat to the continued development of Thai manufacturing, particularly in low-wage, labour-intensive industries such as textiles and garments (Doner and Ramsay 1994). Moreover, it is not only Chinese networks of capital that have increasingly centred their activities on southern China: Japanese investors have also turned to China as a base for offshore production (Steven 1996; UNCTAD 2002), leading to the enormous rise in inward FDI in China.

If regional and global markets were expanding at an adequate rate, the development of manufactured exports from China would pose less of a threat to the manufacturing sector in countries like Thailand. However, the limited development of consumption in Japan (Steven 1996; Bevacqua 1998), exacerbated by the stagnation of the Japanese economy in the 1990s (Bernard 1999: 192), has meant that Thailand remains highly dependent on export markets in the US and Europe, relative to its imports from Japan. For example, from 1990 to 1997, Japan accounted for some 30 per cent of Thai imports but received only 17 per cent of Thai exports, while the United States accounted for only 12 per cent of Thai imports but received 21 per cent of its exports (UN, *International Trade Statistics Yearbook* 1992–2000).

As such, increased competition from China in the US market is a moment of great consequence. China took a larger share of the US import market over the 1990s, increasing from just 3.1 per cent of the total market in 1990 to 7.8 per cent in 1998, while over the same period, by contrast, Thailand's share of the total US market stagnated at 1.4 per cent (UN, *International Trade Statistics Yearbook* 1992–2000; US Bureau of the Census 2000). Moreover, the Chinese state exacerbated the competitive pressures in 1994 when it devalued its currency—a move that occurred in the same year as the North American Free Trade Agreement went into effect, solidifying Mexico's position as a production base for low-wage exports to the United States. In sum, increased global competition, particularly from China, and the difficulties this created for expansion of exports in an export-dependent economy helps explain the overcapacity evident in the Thai manufacturing sector by the mid-1990s. But overcapacity only explains part of the decline in profitability of Thai manufacturing.

6.2.3 The crisis as a profit-squeeze crisis

As shown in Figure 6.2, declining profit share after 1990 also placed downward pressure on profit. What explains this declining profit share? First of all, from at least the early 1990s, business leaders expressed concerns about a tight labour market, especially for skilled labour, which was driving up labour costs (EIU 1992). It has even been proposed that there is somewhat of a shortage of unskilled labour for many manufacturing industries, though this is a phenomenon in need of careful analysis, given the abundance of labour remaining in relatively low-wage agricultural occupations. Aside from the issue of possible labour shortage, another important theme in the 1990s has been a resurgence of labour militancy, registered in increasing numbers of strikes and workdays lost from 1990 to 1997 (Pasuk and Baker 1998). Labour militancy might be seen as reflecting the increased bargaining power of labour under conditions of relative labour shortage, or as an overdue response to years of wage repression, or as a combination of these factors. It can also be seen as part of the emergence of more democratic forces in Thai society, which have pushed for political liberalization and demilitarization (Ji 1997). However interpreted, it is clear that the increase in militancy placed upward pressure on wages in the leading industrial sectors with the largest unions.

Rising wages, however, do not necessarily imply the unfolding of an economic crisis—indeed, for Keynesian theory, rising wages might well help stimulate domestic consumption and thus set in motion the virtuous circle of demand-led investment and economic growth. Rising wages only represent a problem in the context of socially disarticulated and export-led accumulation—particularly if wage increases are not outstripped by productivity growth, in which case rising wages will contribute to immediate declines in the profit rate. Yet capital accumulation in Thailand, as in many other peripheral countries, has not been marked by the kinds of consistent productivity increases that can lead to a virtuous circle of demand-stimulated economic growth—a matter discussed in the next section.

Here it is also important to re-emphasize that the Asian NICs compete with one another in the same major export markets and have limited complementarity, each being dependent upon exporting a large surplus that cannot be consumed in the domestic market.[5] Though Thailand has not been successful in improving its regional competitiveness through productivity gains, it was successful throughout the 1980s in maintaining competitiveness by keeping wage growth well below growth in labour productivity. In the early 1990s, however, rapid wage increases outstripped increases in labour productivity (Figure 6.3). While such wage rises could be beneficial if sustained over time (through their effects on the domestic market), in the short term they have not been sufficient to allow Asian exporters to consume all of what they produce, and have instead simply undermined the price competitiveness of many exports.

6.2.4 The crisis as a crisis of productivity

It remains to be asked why, in this context of increased wage pressure, there was not a comparable, countervailing increase in the productivity of capital. To answer this question, it is important to re-emphasize the historical features of Thailand's dependent development discussed in previous chapters. Industrialization in Thailand has for many years been built heavily around

[5] For comment on this in Thailand, see e.g. *The Nation*, 5 Sept. 1997; *Bangkok Post*, 18 June 1998.

Fig. 6.3. Manufacturing wages and labour productivity, whole Kingdom, 1983–1999.
Sources: Calculated from NESDB, *National Accounts* (1983–99); NSO, *LFS* (February round, 1983–99).

low-wage labour. Real wage growth was extremely limited between the end of World War II and 1975, and was far slower than labour productivity growth from 1975 until 1990. Thus, firms investing in Thailand could enjoy extremely high profit rates with relatively limited investment in productivity enhancement.[6]

This is not to say that technological change and upgrading have not occurred throughout the post-war period. Certainly, more labour in Thailand has been harnessed to new production technologies, and this is reflected in labour productivity increases, which, while not on par with those of the other Asian NICs, are nonetheless significant. Moreover, even total factor productivity—which mainstream economists problematically take to indicate technological upgrading through innovation or imitation—has improved in Thailand over its recent decades of rapid growth according to some accounts (World Bank 1993; Crafts 1999). Nonetheless, as is to be expected for a country engaged in 'catching-up' growth, the major source of productivity change is not innovation but rather simply increased utilization (largely through FDI) of more advanced machinery, along with the transfer of labour out of agriculture into manufacturing (Krugman 1994; Crafts 1999; Warr 1999).

[6] A World Bank survey of firms in nine East and Southeast Asian countries, along with the United States and Germany, found Thailand to have the highest rate of return on assets over the period 1988–96—and also to have the most rapidly declining rates of return over the 1990s (Claessens, Djankov, and Lang 1998).

In this context, a large number of firms in Thailand have remained focused on the advantages to be gained from relatively competitive wages, rather than on investment in increased productivity of capital. Indeed, as already noted, even larger foreign firms in Thailand have made only limited efforts to train personnel and upgrade technology used in their Thai manufacturing operations, a problem that has been exacerbated over the post-World War II period by the policies of the Thai state.[7] The Thai state's relatively poor performance in spurring scientific and technological research and development (R & D) is reflected in the fact that as of the mid-1990s Thailand had only 0.2 R & D scientists and technicians per 1,000 members of the general population—a level not only much lower than that of competitor states such as South Korea (2.9) and Singapore (2.6), or even China (0.6) and Vietnam (0.3), but also lower than the average for all countries categorized by the United Nations Development Program as exhibiting medium human development (0.7) (UNDP 1999: 176–7). To be sure, there have recently been modest increases in the proportion of government spending on education and science and technology development. But in spite of this, productivity of capital, as measured by the capacity–capital ratio, declined between 1982 and 1996—a continuation of the relatively low-productivity pattern that has marked Thai accumulation throughout most of its recent history.

The ability of the Thai state to pursue this comparatively low human capital development path, it should be noted, has roots in the country's abundant land and natural resource base, which has allowed several decades of robust export growth on the basis of primary commodities—primarily rice (Pasuk and Baker 1995; Jomo 1997). But Thailand's development path is also rooted in the effects of the Cold War on the evolution of the Thai state and its forms and functions. Thailand's development under US umbrage gave it various forms of wherewithal to discipline labour while promoting rapid growth that was underpinned originally by substantial aid flows, and later by FDI and favourable terms of access to the US market. This meant that high profit rates and opportunities for economic growth could be maintained for many years without disciplining capital or developing the capacity to force capital to invest more in training and education of labour—e.g. through higher effective rates of corporate taxation to support greater state expenditure on education, training, and technology development. Thailand in fact has one of the lowest rates of effective corporate taxation in the World (Warr and Bhanupong 1996: 75–6), but until the 1990s it countered the problems this poses for the state's capacity to invest in skills upgrading and technology development by maintaining a labour force so disciplined that it had one of the lowest wage shares of any labour force in the developing world (UNIDO 1992: 45; Jansen 1997: 34–5). When this changed in the 1990s, under the impact of tighter labour

[7] A 1998 study found that Thai firms are less productive than regional competitors in places such as Malaysia (*Bangkok Post*, 18 June 1998).

markets and more militant labour struggle, neither the Thai state nor most of the private sector had developed substantial capacity to respond to rising wages with labour-displacing technological innovation. Instead, even though some larger firms invested in 'catching up' technological change, a great number of firms simply ratcheted up investment and production on the basis of existing, labour-intensive technology, attempting to outcompete one another through 'perspiration rather than inspiration' (Crafts 1999; Warr 1999).[8]

Here, it is also worth emphasizing the pace at which the realization and overcapacity crisis has evolved—unfolding largely since the second half of the 1980s—which has meant that countries attempting to maintain themselves as export bases have to respond quickly and harshly to rising production costs. The simple means of doing this is to constantly devalue the currency; the more difficult means is to move rapidly from lower value-added products suffering from increased competition into higher value-added lines that are (somewhat) less competitive. The speed required of the transformation process in the era of 'globalization'—a reflection of what David Harvey calls 'space-time compression' (Harvey 1989)—is important to emphasize because, again, it is not the case that firms in Thailand have *totally* neglected efforts to increase productivity or move into higher value-added lines, nor is it the case that the Thai state has totally neglected efforts to spur industrial deepening (Deyo 1995*a*, *b*; Pasuk and Baker 1998: 43–5). However, in a context of historically satisfactory profits insured by repression of labour and minimal disciplining of capital, investors and state planners have simply been unable to respond with the timeliness or seriousness required by contemporary global capital flows (Doner and Ramsay 1994: 189–94; Bello, Cunningham, and Poh 1998: 55–70).

6.2.5 The crisis as a financial crisis

Realization failure, rising wages, and stagnant productivity in the early 1990s provide a crucial context for understanding the Thai economic meltdown of 1996–7, but they do not by themselves explain the crisis or its onset. The issue of capital switching is also important. As Figure 6.4 shows, one of the most significant features of FDI inflows during the early 1990s is the decline of new investment in manufacturing industries and the explosion of investment in real estate that began in 1994 and continued through 1996. It is also worth noting that this investment in real estate—which quickly led to a glutted market—is not matched by comparable investment in construction, indicating that much of the real estate investment involved speculative land deals (cf. Bello, Cunningham, and Poh 1998: 161). Indeed, the national accounts indicate that

[8] It is worth cautioning here that productivity increases in core countries—to which countries like Thailand have been unfavourably compared—may be overrated in some of the literature, since total factor productivity does not separate out the effects of innovation and imitation from the effects of less efficient firms exiting or more technologically efficient firms gaining market share (Webber and Rigby 1996: 394, 401–2).

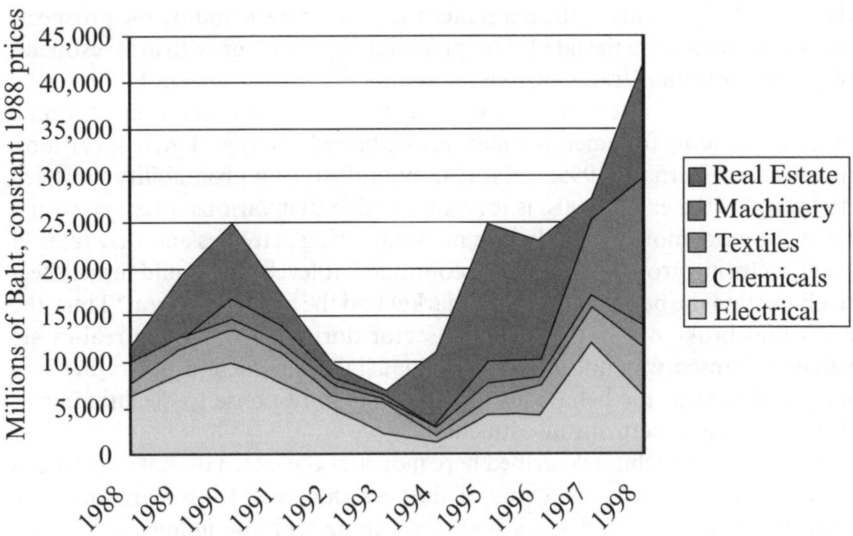

Fig. 6.4. Foreign direct investment, by sector, 1988–1998.
Source: Bank of Thailand, *Monthly Bulletin of Statistics* (1988–98).

compared to the period 1987–91, growth rates of value-added in the manufacturing and construction sectors declined much more during 1991–5 than did rates in banking, insurance, and real estate (NESDB, *National Accounts* 1987–95). Moreover, during the early 1990s portfolio investment overtook FDI, and total stock market capitalization reached a level equal to one-tenth of GDP by 1993 (UNCTAD 1997). All of these indicators point to the development of a much higher level of speculative, non-productive investment by 1993–4. This movement of capital into speculative activities is not entirely surprising, given low or declining profit rates in all sectors except BIR, and booming profit rates in BIR themselves reflect the increase in speculative activity in real estate.

The phenomenon of increasing speculative investment is described in mainstream analyses of the economic crisis as the development of the 'bubble economy'. It is widely acknowledged by analysts that the shift of capital into more speculative ventures not only set up the subsequent crash in real estate values—particularly since such investment was swelling at a rate much faster than the growth of the overall economy, which was slowing during the early 1990s—but also represented a diversion of capital from employment in productive sectors like manufacturing (Pasuk and Baker 1998: 316–17). What needs to be further emphasized, however, is that the timing and weight of the movement of capital into more speculative investments seems quite plausibly to imply a response by investors to declining profit rates in manufacturing by the early 1990s. Firms in the manufacturing sector were finding the prospects less attractive than in the late 1980s, precipitating slower growth in investment, while extraordinary rates of overall economic growth up until 1990—and respectable rates after that—had attracted the attention of institutional investors looking for emerging markets within which to park new short-term investments (Bernard 1999: 191). Had manufacturing profitability not been declining by the early 1990s, is it is conceivable that various investors would have put more money directly into manufacturing activities and that rates of manufacturing growth would have continued at levels that could more effectively sustain the boom in the stock market and the property sector.[9] Thus, the crisis that broke out in the financial sector during 1996–7 when real estate values collapsed was not a narrowly financial phenomenon but rather was directly linked to the behaviour of capitalists in response to declining profitability of manufacturing investment.

The capital switching described here indicates the international dimensions of the issue. It was the switching of East Asian capital from manufacturing FDI into real estate and stock markets—along with the inflow of financial capital from Western banks and emerging market funds (itself a form of switching from reinvestment in the slower-growing core countries)—that triggered the development of Thailand's bubble. To point out these international dimensions of the bubble economy, however, is not to say that international capital simply ran roughshod over the Thai state. Rather, the Thai state was very much an actor in the process, and in order to encourage rapid inflows of foreign capital and bridge a projected savings-investment gap, it had undertaken a number of financial liberalization measures during the early 1990s (Bello 1998; Bello, Cunningham, and Poh 1998: 18–20; Pasuk and Baker 1998: 98, 116–17, 318–19; Unger 1998: 95–7; Bernard 1999: 191). First, the state deregulated domestic finance and removed constraints on portfolio management, including loosening rules on capital adequacy requirements and allowing commercial banks and financial institutions to expand their fields of operation. Second, the state dismantled most foreign exchange controls and opened the Bangkok International Banking Facility, which allowed offshore borrowing in foreign currencies and reconversion into Thai baht, thus increasing the flow of capital into Thailand from countries with lower interest rates. This was further facilitated by the third policy, that of keeping the baht pegged to a weighted basket of currencies favouring the dollar, thus making dollar-denominated loans artificially cheap as the dollar rose against the yen. Fourth, the state kept interest rates high to attract foreign capital.

In sum, while the financial dimensions of the crisis clearly implicate volatile international capital flows, the development of crisis tendencies in the Thai

[9] Pasuk and Baker (1998: 101) note that between 1992 and 1993 the price/earnings ratio of companies listed on the Stock Exchange of Thailand rose from 16 to 26, while that of blue chip companies rose to over 30.

economy indicates a more complex process than signalled by the conventional story of international capital run amok. Aside from the important immediate roles of Thai capital and Thai labour in the development of crisis tendencies, international forces were themselves constituted in part by the actions of Thai capitalists and the Thai state—including their export performance and their financial manœuvring in the post-Cold War process of liberalization. The financial crisis was integrally connected to crisis tendencies within 'the real sector' and was simultaneously produced by 'internal' and 'external' forces.

6.2.6 The crisis as a unity-in-diversity

The different aspects of the economic crisis discussed above are not separable and independent causal forces, but, rather, dialectically interconnected phenomena giving the crisis in Thailand its specific form. Moreover, the crisis is simultaneously a national and international phenomenon; indeed, the Thai economy as a national entity contributed to the creation of many of the regional and international dynamics that have acted back on the accumulation process in Thailand.

To summarize synthetically, then, the crisis in Thailand had roots in a decline in manufacturing profitability. This decline in profitability resulted from the interaction of realization failure (the consequence of overcapacity in a context of increasing export competition), profit squeeze from rising wages (the result of tighter labour markets and increased worker militancy), and declining productivity of capital (the result of limited capacity for technological upgrading, which is the legacy of dependency and specific historical factors). The growth boom that started in 1987 had in effect quickly led to over-investment without adequate productivity growth. In this context, rising wages insured declining profit rates while further exacerbating the problem of competition from lower-wage competitors.

These manifestations of underlying crisis tendencies enabled a full-blown meltdown in the financial sector, triggered by the collapse of a speculative economic bubble built up through large flows of both foreign and domestic capital into the real estate sector and the stock market. The state facilitated both the original boom in manufacturing FDI that led to record growth rates during 1988–90 and the dramatic inflow of hot money during 1993–4 as the manufacturing sector cooled somewhat and investors began to look elsewhere. The phase of open crisis began with the collapse of real estate values and manufacturing export growth during 1996, which led to disinvestment from the stock market and speculative pressure on the baht—many investors having decided that the currency was in need of devaluation in order for exports to regain competitiveness.

From this point, the crisis began to take on a regional life of its own, which I will not analyse here (see Glassman 2003). One concluding point about the regional dimensions of the Asian crisis is in order, however. None of the countries affected by the regional crisis had Thailand's precise mix of liabilities, and this, coupled with the fact that crisis in other countries was clearly triggered—though not necessarily caused, in any narrow sense—by the herd behaviour of panicked investors, has meant that financial contagion effects emerged as the predominant focus of attention (e.g. Jomo 1998). While there can be little doubt that financial contagion did in fact ignite the meltdown of currencies and stock markets throughout Asia, the analysis here suggests the need to examine the bases of the regional crisis in problems of overcapacity and declining profitability. Given this context, it is entirely possible that a regional crisis could have been triggered by a downturn elsewhere in the region, though perhaps not on the same timetable or in precisely the same fashion.

6.3 Economic crisis and struggle over the Thai state

Though the economic downturn in Thailand took hold in 1996, during the short tenure of a somewhat less internationalist regime headed by former military commander Chaovalit (1996–7), the analysis presented above shows that the crisis was in fact a process in motion throughout the 1990s and thus was at least as much the product of activities undertaken by previous, more internationalist regimes, including the highly internationalist Democrat Party regime of Chuan (1992–5). Nonetheless, the crisis created an opportunity for the Democrats to reclaim power, with strong backing from both Bangkok supporters and international business. Chuan regained the Prime Minister's position in 1997 and subsequently held this position for three years while presiding over implementation of the IMF's SAP. Thus, the Democrats became the party most strongly identified with neo-liberalism in Thailand (Pasuk and Baker 2000).

It is worth briefly noting some of the major features and consequences of the SAP, which did much to set the stage for subsequent political manœuvrings. Along with mandating exchange rate flexibility, Thailand's SAP originally emphasized cuts in central budget expenditures (even though debts were overwhelmingly held by the private sector), with a targeted budget surplus equal to 1 per cent of GDP for 1997–8 (IMF 1997). Capital inflows were initially to be encouraged through high interest rates and eased restrictions on equity participation in troubled financial institutions. Restructuring of the financial sector included the closure of 58 insolvent finance companies. Wage increases were to be pegged to inflation, though in actuality the purchasing power of the minimum wage fell. The state also announced its intention to encourage privatization of state enterprises in the energy, transportation, utility, and communications sectors (Lauridsen 1998).

As the economic situation in Thailand worsened throughout 1998, with GDP declining by 10.5 per cent (Bank of Thailand, *Annual Report* 1999), the

Chuan regime negotiated some changes with the IMF. High interest rates, that encouraged a continuing sense of crisis amongst foreign investors and crushed many local businesses, were slowly lowered, and the state was allowed to run a budget deficit equivalent to 5 per cent of GDP during 1998–9 (IMF 1998). These reflationary measures helped the economy, with GDP growth recovering in 1999 and 2000 to 4.4 per cent and 4.6 per cent respectively (Bank of Thailand, *Annual Report* 2000).

More direct measures ensured that certain private interests would be bailed out with public money. IMF funds were used to pay off the central bank's obligations and thus to indemnify foreign investors, as well as to restore currency reserves which had been depleted, in part, by efforts to bail out insolvent local finance companies (Bullard, Bello, and Malhotra 1998). Overall however, these measures were insufficient to save many domestic capitalists—to the benefit of foreign investors who have been able to buy up Thai assets at bargain prices (Hewison 2000). Nonetheless, certain well-positioned Thai elites have also been able to take advantage of opportunities presented by the crisis through activities ranging from arbitrage to new joint ventures with foreign investors.

While some of the edges were taken off Thailand's SAP to facilitate the restructuring of capital, there were fewer efforts to directly rescue others. As the SAP took hold, unemployment more than doubled—from 1.9 per cent of the workforce in 1997 to 4.2 per cent in 1999, according to the Bank of Thailand (Bank of Thailand, *Annual Report* 1997–9). Other estimates place the 1999 rate even higher at 5.1 per cent, while estimating a loss of 1.4 million construction and 140,000 manufacturing jobs between 1997 and 1999 (NSO, *LFS* 1997–9). With devaluation and inflation, real wages for manufacturing workers fell from US $188/month in 1996 to US $133/month in 1999 (calculated from NSO, *LFS* 1996–9). Unemployment insurance has not yet been fully developed, and a major programme of poverty alleviation was not put in place until 1999 (Sauwalak 1999). Thus the Thai state under the Democrat Party's structural adjustment regime primarily relied on rural society to act as a shock absorber by finding work and residence for those laid off from urban-industrial occupations. This pushed the rural poverty rate up from 14.9 per cent of the population in 1996 to 21.5 per cent in 1999, accounting for a disproportionate share of the total national increase in poverty from 11.4 to 15.9 per cent (Mingsarn 2003: 87).

A politically significant consequence of all of this has been that as the Thai economy muddled through to semi-recovery, and as various social groups have struggled to deal with the consequences of both crisis and structural adjustment, the Democrats came to be widely identified as having presided over failed and inappropriate policies. This view was reflected in the elections held in early 2001, which brought into power a coalition proclaiming more nationalist and populist priorities, headed by multimillionaire businessman and political insider Thaksin Shinawatra. While there is no doubt that this political outcome reflects a real reconsideration of approaches to development—and not only by popular organizations but also by Thai business leaders who have suffered in their collision with international capitalism and its crisis tendencies (Hewison 2000; 2003)—it can be questioned how nationalist this new coalition either intends to be or can afford to be. In the next section, I briefly analyse the social basis of the Thai Rak Thai-led government and suggest it is a coalition already manifesting the internationalization of the state and is very unlikely to transform development policy in Thailand into anything resembling a more strongly nationalist approach. Nonetheless, the rise of Thai Rak Thai also manifests significant and complex class (-relevant) struggles over the response to both crisis and the broader problems of uneven development and industrial transformation, and in this sense it is important to interrogate both what has changed and what has remained the same as internationalization processes in Thailand have butted up against the limits to capital.

6.3.1 Thai Rak Thai economic policy: nationalism for a post-nationalist era

In the run-up to the 2001 election and its immediate aftermath, Thaksin and Thai Rak Thai promised a series of specific economic policy measures that received much domestic and international attention and which can be taken to constitute the core of Thai Rak Thai 'nationalism'. The specific measures included the following:

(1) Development of the state-run Thai Asset Management Corporation (TAMC) to restructure the debts of commercial banks and relieve them of some of their non-performing loans (NPLs);
(2) Establishment of a 58 billion baht emergency fund to assist financially strapped firms in labour-intensive industries;
(3) Programmes to promote small and medium enterprises (SMEs), including the development of a government office of SME Promotion;
(4) A state-promoted One District (*tambon*), One Product (OTOP) scheme to encourage entrepreneurialism and small business development in rural and upcountry areas of Thailand;
(5) Promises to expand Internet access in rural areas for both educational and entrepreneurial purposes;
(6) A state fund, established from the Government Savings Bank, to provide one million baht to each of Thailand's 70,000 plus villages for a specific development project, to be developed by villagers and approved by a government committee;
(7) A three-year debt moratorium for farmers owing US$2,000 or less to the BAAC;
(8) Promises to negotiate with the AoP over contested energy and natural resource policies that have negatively affected the lives of many villagers;

(9) Development of a national health insurance programme to provide basic clinical services to the uninsured at the cost of 30 baht per office visit.

Each of these projects has a specific kind of 'nationalist' edge that draws support from different groups and helps highlight some of the tensions and contradictions of Thai Rak Thai nationalism. I will look here at five such groups: large, internationally mobile Thai capitalists; owners of small and medium enterprises; state bureaucrats and Thai Rak Thai party members; agricultural and village organizations; and organized labour.

6.3.1.1 Nationalism among the Thai economic elite?

A crucial feature of the emergence of Thai Rak Thai economic nationalism has been support for specific economic restructuring measures by the most powerful Thai capitalists. These include such well-known capitalist family patriarchs as Dhanin Chearavanont, head of the huge CP conglomerate, whose agribusiness and other investments in China had at one time made CP Group that country's single biggest foreign investor.[10] Heads of major Thai banking families whose banks have substantial overseas operations, including Chatri Sophonpanich (Bangkok Bank), have also supported specific economic restructuring measures that have been pushed by Thai Rak Thai.[11] Here, I look briefly at the support given by Dhanin and Chatri to Thai Rak Thai, taking their reasons as indicative of the reasons that a variety of other major Thai capitalists supported Thaksin's election bid and specific economic policies.

For those who consider Thai Rak Thai's orientation to be nationalist in any very conventional sense, elite capitalist support for Thai Rak Thai might seem a surprising development, since these Thai capitalists have been very regionally and internationally mobile, and seem to have developed a strong commitment to participating in processes of 'globalization'. Yet such support can also be seen as a conventional reaction to the effects of economic crisis. The majority of even the most powerful Thai capitalists were negatively affected by the crisis—some devastatingly so (Hewison 2000; 2003). Dhanin's estimated net wealth declined from US$5.5 billion in 1995 to US$4.7 billion in 1996 and US$1.7 billion in 1997, before he dropped out of the billionaire club entirely by the year 1998. Chatri saw his net wealth decline from US$3 billion in 1995 to US$2.2 billion in 1997 and less than one billion in 1998.[12]

The scale politics of capitalists such as Dhanin and Chatri can be interpreted as a response to their very real and increasing difficulties in a more intensely competitive international environment and their need for specific forms of support from the only major political organization they can hope to rally to the cause of providing assistance—the Thai state. The CP Group, for example, announced during the crisis that it would sell off many of its non-core operations and focus on consolidating its position within its core areas of business

[10] *Bangkok Post*, 31 Oct. 2000. [11] Ibid. 29 Nov. 2000.
[12] Ibid. 22 June 1998; 23 June 2001.

strength.[13] This involved attempting to shore up core agribusiness operations in Thailand.[14] At the same time, CP Group's newest and perhaps most important line is its telecommunications investments, and here it must compete for favours from the Thai state, which plays a crucial role in the success or failure of given firms through its decisions on licences, contracts, ownership regulations, and the like. This competition for the attention of the state has been intensified by Thaksin's role in recent Thai governments, including his position as a minister within the New Aspiration Party government of Chaovalit, since the Shinawatra family's telecommunications empire is one of the main competitors CP Group must confront in its bid to gain a larger share of the Thai market (Baker 2004).[15]

Dhanin has for years subscribed to views that are in any event conducive to support for some of the kinds of policies that Thai Rak Thai has put forward in the name of economic reconstruction. As the crisis descended upon Thailand in 1997, Dhanin—who is reputedly respected and listened to by policy-makers—spoke repeatedly and fervently about the desirability of the Thai government focusing on agriculture and attempting to lift agricultural incomes as a way to rebuild the economy.[16] CP group has been particularly interested in encouraging increased output of rice exports, and established a rice mill in Buri Ram province as a model of provincial enterprise that it has advertised to both the government and the private sector.[17] Interestingly, the estimated cost of investment in one such mill is 1 million baht, and though the sources of the Thai Rak Thai Party's inspiration for its much-vaunted 1 million baht per village development scheme are no doubt varied, it seems likely that the sort of thinking that Dhanin has advertised about how to stimulate growth in rural Thailand has filtered through to people like Thaksin. This is also suggested by Dhanin's advocacy for getting more farmers hooked up to the internet[18]—a project that conveniently favours both of CP group's core businesses and also happens to have become a pet idea of the new prime minister, whose telecommunication interests could also be well served by such a plan (Crispin 2001).

Similar kinds of considerations apply when considering the support given to Thai Rak Thai by the large Thai banking families. Thai banks enjoyed a long period of protection from foreign competition (1963–89) because of state policies restricting new licences for foreign banks (Hewison, 1989*a*). These protections were gradually lifted during the 1990s economic boom, and in the wake of the crisis, with its devastating effects on many banks and finance companies, foreign investors have played an increasing role and have been encouraged to help recapitalize the financial sector. While some bankers may welcome the increasing foreign presence as the only way to shore up an ailing

[13] *Bangkok Post*, 1 July 1999; 4 Jan. 2000.
[14] Ibid. 6 Nov. 1998.
[15] Ibid. 15 June 1998.
[16] Ibid. 5 Sept. 1997; 6 Nov. 1998; 12 Mar. 1999, 19 May 2000.
[17] Ibid. 4 Jan. 2000; 19 May 2000.
[18] Ibid. 20 Jan. 2000.

system, there can be little doubt that these new developments threaten a real and perhaps permanent loss of power—symbolized acutely by the formerly powerful Tejapaibul family's loss of control of the Bangkok Metropolitan Bank (Hewison 2000: 203).

Thus, it is not surprising to see powerful banking families attempting to shore up their positions through state policies. Chatri, in particular, has spoken favourably of the TAMC, and it is clear that if it is structured and developed in the right way it could assist banks like Bangkok Bank, shifting the burden of restructuring their NPLs to the state.[19] The TAMC has been an object of some contention, with some bankers being lukewarm about it and various economists taking different views of its merits and appropriate structure (Kasian, 2002; Pasuk and Baker 2002: 9). By supporting Thai Rak Thai and throwing its weight behind the TAMC, however, Chatri may have hoped to get in on the ground floor in helping craft policies that would strengthen the position of banks like his own. By the end of 2001, Thai Rak Thai's first year in office, Chatri's support seems to have paid dividends, with Bangkok Bank being reported to have had some 60 billion baht worth of its NPLs transferred to the TAMC, more than any other private bank.[20]

The kind of support given by investors like Dhanin and Chatri to Thai Rak Thai policies is based fairly clearly on specific interests, and is in many respects more narrowly neo-mercantilist than broadly nationalist. At the same time as they have favoured specific state interventions that would shore up their own industries, they have also spoken out in favour of foreign investment in general and have clearly wished to maintain Thailand's strongly export-oriented economic profile.[21] Moreover, Chatri and the Bangkok Bank played an important role in forcing a change of management at the Thai Petrochemical Industry (TPI), a change in which the founding Thai owner-manager, Prachai Leopairatana, was replaced by an Australian management company, arousing an opportunistic nationalist response from the deposed owner.[22] In this sense, the nationalism of Thai economic elites, like that of Thai Rak Thai itself, is very unlike the strong import substitution-oriented nationalism of Latin American countries during the 1950s and 1960s, and at best seems an attempt to gain the kinds of specific statist supports for domestic industry that characterized South Korea during its post-World War II period of neo-mercantilist economic policy.

The support given to Thai Rak Thai by various Thai business elites thus manifests the real crisis afflicting Thai capital, in relation to which these capitalists are scaling down their operations in certain areas and attempting to reconsolidate their positions through concentrating on investments in core businesses while cultivating crucial forms of state support. Such activities are

[19] Ibid. 5 Sept. 2000; 3 Feb. 2001.

[20] *Bangkok Post Year-End Economic Review* 2001.

[21] *Bangkok Post*, 9 Sept. 1997; 30 Sept. 1997; 28 Nov. 1998.

[22] Ibid. 16 Mar. 2000; 17 Nov. 2000; 28 Dec. 2000; 25 Sept. 2001; 26 Oct. 2001.

entirely routine forms of business manœuvring. The issue of whether or not they form the basis for any deeper nationalist agendas is one that depends upon both the specific context in which these manœuvres are carried out and how successful they are. To explicate that context further requires examining the basis of support for economic nationalism among other social groups in Thailand.

6.3.1.2 Nationalism and Thai small and medium enterprises (SMEs)

While support for Thai Rak Thai economic policies among the much larger and more heterogeneous group of SME owners of the country is much harder to gauge than among the Thai economic elite, there is evidence that such enterprise owners have favoured Thaksin's agenda on the same sort of opportunistic basis. During the election campaign, Thai Rak Thai made much of how it was going to focus on smaller business interests that had been neglected in the Democrat Party's monolithic focus on rebuilding the tattered finance and banking sectors. Various promises were made to provide state support for SMEs through new lending facilities and other programmes. These promises seem eventually to have been met with approval by business groups throughout the country, and Thaksin's portrayal of himself as someone who started as a small entrepreneur and thus understands the needs of these kinds of businesses helped to enhance that approval (Pasuk and Baker 2002: 5–6).[23]

Reflective of the way in which Thai Rak Thai gained the support of these SMEs was its establishment of the 58 billion baht emergency relief fund for labour-intensive industries. While this is not a fund available only to SMEs, many financially strapped SMEs in labour-intensive industries such as garment, textile, and footwear production would be among the most likely recipients. While a large number of such potential recipients are clustered in the Bangkok area, moreover, provincial Chambers of Commerce have lobbied Thai Rak Thai to set aside 10 per cent of these funds for provincial businesses, a larger percentage of which are SMEs.[24]

Economic policies like the emergency relief fund for labour-intensive industries and the OTOP scheme, along with the forms of state intervention favoured by the bigger Thai capitalists, are clearly motivated less by a comprehensive sense of nationalism than by the narrow needs of hard-hit smaller firms. Since many are exporters (investment-constrained capitalists), the state funds into which they can now tap will not in fact produce a more insular economy but will rather, at best, allow them to continue competing for a while longer in export markets that have become increasingly competitive. Thus, the emergency relief fund for labour-intensive industries is, like the other policies mentioned above, a means for the Thai state to promote a form of

[23] Author interviews, business people in Chai Nat province (Jan. 2001), Ubon Ratchathani province (25 Dec. 2001).

[24] *Bangkok Post*, 19 Nov. 2001.

neo-mercantilism favoured by a particular segment of the capitalist class (Pasuk and Baker 2002; Baker 2004). This consideration also applies to the 1 million baht per village development fund, also supported by many provincial businesspeople for its potential inflationary effects on the provincial economy.[25]

6.3.1.3 Nationalism and the Thai state

If capitalists clearly have interests that transcend state boundaries, state officials who are not also capitalists are generally thought to have interests that are more territorially bounded, though I have argued that this view can be erroneous. Nonetheless, insofar as state officials have interests in national programmes that demand state funding, and in facilitating the kinds of growth within state boundaries that can generate the tax revenues for maintaining these programmes, there are surely times and places that political leaders have a strong interest in nationalism.

Yet in assessing the role of the Thai state in relationship to Thai Rak Thai nationalism, the picture appears somewhat more complicated. It is important here to differentiate between the permanent bureaucracy and elected officials, as well as between those elements of the bureaucracy that have become strongly internationalized and those that have not. For example, although some managers of organizations such as state enterprises have favoured privatization, others have resisted such neo-liberal measures, and one of the most outspoken opponents of privatization and foreign ownership of state enterprises has been the former managing director of the state-owned Bangchak Petroleum Company, Senator Sophon Suphapong.[26] At the same time, many bureaucratic structures have become highly internationalized over the years through ongoing collaboration with international lending agencies and the representatives of other states. Thus, many of these organizations—such as the Electricity Generating Authority of Thailand, which has resisted various attempts by Thai Rak Thai officials to make it compromise with the AoP over its projects (Glassman 2002)—are capable of standing in opposition to policies of the elected officials.

Ultimately, in assessing Thai Rak Thai nationalism, one must look carefully at the party itself. Thai Rak Thai is first and foremost the electoral vehicle put together by Thaksin out of his own ambitions. Given Thaksin's enormous wealth, estimated at US$1.2 billion or more (Hewison 2003: 7; Pasuk and Baker forthcoming),[27] and his international business interests, this makes Thai Rak Thai a seemingly poor candidate for a party devoted to serious forms of economic nationalism. In Thaksin's case, it is important to foreground the fact that he has always been a capitalist who depended on different state contracts

[25] Author interviews, members of Ubon Ratchathani province Chamber of Commerce (25 Dec. 2001).

[26] *Bangkok Post*, 5 May 1998; 22 June 1998; 15 Aug. 1999.

[27] Ibid. 23 June 2001.

and regulatory decisions for his projects. Thaksin got his start in the telecommunications business, for example, through a contract to supply computer software for the Thai police,[28] and has long ensconced himself in governmental positions in ways that contributed to development of his company's fortunes. Thus, he is far less likely to be doctrinaire about the role of the state in the economy than some political figures, and in this sense he can be seen—very much like Dhanin and Chartri—as a capitalist who is willing to use the institutions of the national state to promote internationalist projects.

At the same time, it is important to recognize that Thaksin has long surrounded himself with a variety of different types of advisers, and featuring prominently among these have been a significant number of former Leftists and progressive activists, primarily from among the 'Octobrists', as they are dubbed by Thai scholar Kasian Tejapira (Kasian 2002)—namely, the former student activists prominent in the uprisings against the Thai military dictatorship in October 1973 and October 1976. This 'Octobrist' presence in Thai Rak Thai lent to the party some of its original populist aura, the 'Octobrists' being associated with the struggles of the 1970s to strengthen the positions of peasants and industrial workers. Moreover, this populist aura is also implicitly nationalist in the sense that the projects the 'Octobrists' were associated with in the 1970s were strongly anti-imperialist and devoted to strengthening Thai civil society as part of the struggle for political change. This populist-nationalist aura was particularly powerful as Thai Rak Thai moved into office under the mandate to rebuild an economy many Thais believed to have been ravaged by international forces.

Here too, the reality behind the image is somewhat complex, particularly given splits among the 'Octobrists', but perhaps the main point worth noting is the difference between the projects that attach to this strand of nationalism as compared to those that attach to the neo-mercantile nationalism of the business elites. The populist and nationalist faction within Thai Rak Thai might be seen as the faction that most represents a project placing the interests of the Thai state ahead of business per se, even if this is primarily a matter of strengthening the institutions within which the politicians make their living. But this scale politics, it should be noted, does not imply closed-door economic policies or even policies of bailing out Thai entrepreneurs. Rather, it centres around a sort of national *political* entrepreneurship, and as such, it is focused on the ways in which specific state policies can strengthen bases of political, economic, and electoral support among particular segments of the polity.

6.3.1.4 Nationalism and Thai village organizations

Thai villages are increasingly heterogeneous in their social structures and types of economic activities. I cannot pretend to present here a comprehensive account of the kinds of social groups that exist in villages or the kinds of

[28] *Bangkok Post*, 22 Mar. 2001.

specific interests they have that might lead to support for specific Thai Rak Thai policies, but I will indicate two important types of organizations that Thai Rak Thai has clearly attempted to appeal to in its campaign promises and subsequent programmes.

The first of these are farmer and agricultural organizations, including crop price support groups. Though such organizations did not play a formative role in the development of the Thai Rak Thai party or its policies, the fact that they can potentially claim to represent a huge proportion of the nation's voters means that they have in fact exercised an indirect influence on Thai Rak Thai (Pasuk and Baker 2002: 6–7; Baker 2004). The debt moratorium and the 1 million baht village development fund clearly represent the major initiatives of Thai Rak Thai to generate support among these organizations. The entrepreneurial orientation of some of the larger and more powerful crop price support groups is quite consistent with the entrepreneurial emphasis that Thai Rak Thai has given the village development fund, and so, even though it was a proposal unsolicited by these groups, it has been welcomed.

The response of the second type of village organization, less powerful agricultural groups such as those allied with the AoP (Baker 2000), has been far more tepid. For example, representatives of organizations such as the Northern Farmers' Network have tended to regard Thai Rak Thai's policy pronouncements towards agriculture as little more than an electoral ploy, noting the difference between what was promised in the campaign and what was actually delivered once Thai Rak Thai was in office. Moreover, such groups note that Thai Rak Thai's policies did not percolate up from the demands of organizations such as theirs and thus are not based on a considered approach to the needs of small-scale farmers or poor villagers. As such, they see measures like the 1 million baht per village development fund as means by which the Thai state can attempt to intensify capitalist development in rural areas, but without necessarily helping the neediest members of agrarian society. Certainly, most do not consider their interests to be the same as those of major Thai Rak Thai backers like Dhanin and the CP Group.[29] These suspicions of Thai Rak Thai intentions were heightened during 2002 when relations between the government and Northern agrarian organizations broke down over land invasions in Lamphun.[30] More broadly, relations between Thai Rak Thai and the AoP soured during 2002–3 when Thaksin pulled back from promises to meet AoP concerns over a variety of energy projects that threatened rural livelihoods (Pasuk and Baker forthcoming).[31]

[29] Author interview, representative of Northern Farmers' Network, Chiang Mai province (18 Dec. 2001).

[30] These conflicts were reported in the *Bangkok Post* during July 2002.

[31] The breakdown of relations between Thai Rak Thai and the AoP over the Pak Moon dam is reported in the *Bangkok Post*, throughout January 2003.

6.3.1.5 *Nationalism and Thai workers*

A final social group whose support for Thai Rak Thai policies is worth examining is Thailand's urban-industrial labour force. As with rural and agrarian organizations, this group is heterogeneous and difficult to slot into any simple politics. Thai labour union leaders had endorsed the New Aspiration Party of former Prime Minister Chaovalit during the 2000 campaign, but labour unions cannot deliver the votes of their members. Thus, no one is sure precisely how Bangkok workers voted, but several labour activists to whom I spoke felt that a majority of workers voted for Thai Rak Thai.[32]

For those workers who support Thai Rak Thai, there are a number of different reasons for the affinity. Bangkok workers were hit very hard by the economic crisis, and clearly dissatisfaction with the Democrat Party and the IMF structural adjustment programme it implemented pushed many workers in the direction of what seemed the most viable alternative. Among state enterprise workers, this basis of support for Thai Rak Thai has also issued forth in strong calls to reject many of the privatization measures that have been part of structural adjustment (Glassman 2002: 518). Among private sector employees, there has also been some expression of what could be interpreted as populist nationalism. In one case, that of TPI, workers at one point sided with deposed owner Prachai Leopairatana in protesting against the restructuring plan laid out for the company by its creditors—including major Thai banks—as well as the role of an Australian management company in overseeing that restructuring.[33] Thus, in this instance, there was a confluence of interests between vulnerable workers and specific downwardly mobile Thai capitalists.

But perhaps the most significant Thai Rak policy attracting worker support has been the 30-baht health scheme, which brings access to medical services within reach of many workers who previously could not readily have access to them (Hewison 2003: 13–14). The health insurance programme, it is worth noting, was the brainchild of former Leftists within the Thai Ministry of Health.[34] Thus, the case of the Thai Rak Thai health programme and worker support for it illustrates particularly well the populist face of nationalist politics.

Finally, it is also important to note that Thai Rak Thai has gained some support from labour groups through pushing for the Labour Ministry to make revisions to Thailand's Occupational Safety and Health regulations—in particular, promoting the revisions that had been demanded by Thai labour organizations and activists within the AoP from the time of the 1997 rally discussed in Chapter 3, above. During 1998, labour activists had gained the 50,000-signature petition required by the new constitution to change legislation in support of these revisions, but a deadlock developed when the Labour

[32] Author interviews, Bangkok labour organizers (13–14 Dec. 2001).

[33] *Bangkok Post*, 16 Nov. 2000; 17 Nov. 2000; 29 Nov. 2000; 30 Nov. 2000; 12 Dec. 2000.

[34] Author interviews, Thai Rak Thai officials (Dec. 2001).

Ministry under Chuan's government attempted to forestall such changes by promoting a new labour bill that failed to fully transform the existing administration of the occupational safety and health programme (Chandler and Thong-ek 1998). Thai Rak Thai officials and their New Aspiration Party allies were more willing to court labour groups in this matter and promised to begin implementing the revisions upon taking office in 2001,[35] though little was actually done to honour Thai Rak Thai's promises to labour during its first two years in office (Brown, Bundit, and Hewison 2002: 33).

6.3.2 Many nationalisms; a post-nationalist era

The ways in which the multiple forms of Thai Rak Thai nationalism surveyed here interact—and sometimes conflict—speaks to the heterogeneous social bases and scale politics that lie behind the party's ascendancy. This heterogeneity belies any simplistic identification of Thailand's economic policies as narrowly or conventionally nationalist.

Moreover, a careful look at what economic or other forms of nationalism could plausibly mean for different members of Thai Rak Thai's support base indicates the degree to which any sort of thoroughgoing nationalism has become an impossibility in a post-nationalist era of highly internationalized economies and state structures. Supporters of Thai Rak Thai like Dhanin and Chartri—like Thaksin himself—are in fact among the most internationalized of Thailand's capitalists, and their support for Thai Rak Thai's agenda constitutes little more than standard neo-mercantilist opportunism. Much the same can likely be said of the basis for support of Thai Rak Thai among investment-constrained SME owners. Indeed, few if any Thai capitalists are likely to envisage themselves in a highly nationalist future, given the Thai economy's strong export orientation and dependence on imports, and Thai Rak Thai policies towards capital in fact involve little more than selective interventions to bail out and rebuild troubled firms and sectors.[36] Indeed, even the opportunistic nationalist, Prachai Leopairatana, cuts a sorry figure in this respect, given that he has long been seen as a force pushing for the opening of Thailand's petrochemical industry to selective investment by foreigners (Lauridsen 1999).

The scale politics of other Thai Rak Thai supporters, including those of the party's large rural and agrarian backing and its urban labour supporters, are for the most part 'nationalism' from a different direction than the Thai capitalists. While ailing Thai capitalists are in some respects 'scaling down' and reconsolidating their support within the Thai state, the groups that represent the populist-nationalist dimension of Thai Rak Thai are attempting to 'scale up' to the national level and gain forms of support within the state that they

[35] Author interview, Dr Voravidh Charoenloet, President of the Arom Pongpangan Foundation and Economist, Faculty of Economics, Chulalongkorn University (26 Dec. 2001).

[36] Thai Rak Thai policies have done little to prejudice the basic interests of foreign investors. See e.g. *Bangkok Post*, 12 Jan. 2001; 3 May 2001; 26 Jan. 2002; 29 Jan. 2002.

have not previously had. Some of these groups, such as the AoP or the many private sector unions, are substantially antagonistic to the projects of Thai capitalists (Kasian 2002), and the tensions this creates for Thai Rak Thai are manifest both in factional jostling for position within the party and in Thaksin's increasing hostility towards popular organizations during his second year in office (Pasuk and Baker forthcoming). Thus, even insofar as there is a form of economic nationalism being built by Thai Rak Thai, it is not one but many, and the different forms of this nationalism do not speak to a systematic or strongly nationalist project, let alone to the emergence of a strong and unified developmental state.

6.4 Conclusion

The economic crisis in Thailand was not an event that could be predicted, in any narrow sense, but nonetheless the crisis tendencies at work, which actively contributed to the meltdown of 1996–7, were discernible by the early 1990s. Capitalist economies are never crisis-free and only the shallow boosterism that has shaped much discussion of the Asian NICs allowed the tendencies that lead towards crisis to be ignored. Indeed, just as the socio-spatial unevenness of Thai industrial development has been integrally related to the rhythms and patterns of its relative success, so too the specific features of economic crisis that have emerged are integrally related to the factors leading to the economic boom in the previous period. Rapid growth and burgeoning inequality, boom and bust—these phenomena are not accidentally or contingently related to one another in Thai industrialization but rather are structurally intertwined as part of the contradictory totality that is (peripheral) capitalist development.

The high growth economy of the mid-1980s to mid-1990s was predicated on hitching the 'indigenous' forces conducive to economic growth to international forces that could spur much higher rates of growth than would otherwise be possible. In particular, connecting a socially and sectorally disarticulated 'domestic' accumulation process to the global economy (i.e. internationalization of capital and the state) led to a much more rapid rate of economic and industrial growth than could have been sustained on the basis of limited domestic demand. But this very dependence on global capital flows and markets meant that when the global context for growth became less favourable the internationalization processes that had seemed growth-enhancing instead amplified the inevitable cyclical downturn of the 'domestic' economy and triggered a much more volatile process of asset devaluation than might have otherwise occurred. In short, the Thai state's active embrace of neo-liberalism fostered rapid growth when international circumstances were favourable and rapid collapse when those same circumstances turned less favourable.

To put the issue this way, then, suggests why neither the booster's optimism nor the more dour pessimism of early underdevelopment approaches is adequate to examination of capitalist crisis in Thailand. The geographical, historical, and social unevenness of capitalist development—a deeply rooted feature of its most basic dynamics—ensures that there are a variety of possible trajectories that can be actualized in any given situation. In a case of peripheral development such as that of Thailand, the condition of being peripheral will necessarily influence the specific course of events. This influence could include fostering a more *derigiste* approach to development on the part of the state (e.g. South Korea) in order to counter some of the features of dependency, or it could include a somewhat more complete capitulation to the demands of core powers (e.g. Thailand). These specific strategies will affect outcomes, but unless they eliminate the condition of dependency, the general pressures that capital accumulation in the core create will still be keenly felt, as evidenced by the shared economic plight of such different states as those of Indonesia, South Korea, and Thailand (Glassman and Carmody 2001; Glassman 2003). Indeed, dependency has still characterized all the Asian NICs in the 1990s, and changes in global capitalism (particularly in China, Japan, and the United States) have deeply affected their ability to collectively succeed in the uncomplementary, socially disarticulated, export-led growth projects of the past. The futures of these 'dependently developing' countries will not be predictable, but they will surely continue to be marked by the unevenness and social struggle that are integral to capitalism—and insofar as the Asian NICs remain in varying degrees of dependency, their ability to contain the effects of international forces on their rhythms of development will be limited.

These limitations mark the emergence of the Thai Rak Thai economic recovery agenda. Scarred by the crisis and the incomplete process of recovery from it, much of the Thai population—and even leading segments of the capitalist class—have begun to look with a much more jaundiced eye upon the policy recommendations of neo-liberal institutions like the IMF. In this context, it appears on the surface as if three years of IMF-mandated structural adjustment under the Democrat Party may have generated a response that challenges deep-seated tendencies towards internationalization of the state. Yet below the surface of Thai Rak Thai rhetoric, it is apparent that the leading actors within the state remain not only committed to internationalization of capital but structurally dependent upon it. Nationalism in the Thai Rak Thai mode is unlikely to extend as far as a thoroughgoing Keynesianism or an increased emphasis on ISI policies. Rather, Thai Rak Thai nationalism is a complex and sometimes contradictory mix of specifically targeted policies to prop up the fortunes of specific constituencies. Whether these policies work or not, it is unlikely that they will by themselves lead to any derailing of the broader tendencies towards internationalization of the Thai economy. Moreover, with these policies being promoted by one of the richest and most internationalized business leaders in Thailand, there is every reason to believe that they will be used precisely to try to strengthen the terms on which Thai capitalists engage with the broader regional and global economies, not to promote greater economic autonomy.

7

Conclusion: Thailand at the Margins

The economic crisis and the rise of the Thai Rak Thai Party supplies a paradoxical dénouement to a half-century of rapid Thai economic growth and industrial transformation—as well as to a much longer period of internationalization and general social change. The paradox, however, is not an unsolvable riddle but rather the contradictory character of dependent capitalist development, which inevitably brings destruction along in the train of creation, predicating new possibilities of accumulation on processes of violent devaluation. The Thai political economy now experiences this volatility in virtually full force, having been 'opened' and integrated into the rhythms of global capital accumulation over the course of more than a century. The seeming stability and predictable growth of the years between 1950 and 1995 were facilitated tremendously by Thailand's integral role in the Cold War system, which created various 'conjunctural' cushions against the underlying volatility (e.g. US aid, favoured trade status). With the end of the Cold War system, the Thai political economy is now being increasingly thrust into the less predictable world of global neo-liberalism and post-Fordism and is thus less cushioned against capitalism's 'gales of creative destruction'.

To say this is not to say that growth and industrialization are now on hold. The crisis may well open new opportunities for accumulation and even resumption of rapid growth. But even successful capitalist growth has always done damage to a significant portion of the population—creating and perpetuating enormous socio-spatial disparities—and it will very likely continue to do so in the future. At the same time, the vagaries of the era of 'globalization' may make the growth dynamic much rockier than in the past. Boosters of East Asian 'miracles' should be reminded that other political economies—e.g. Mexico, Brazil, Argentina—have been regarded as 'miracles' in the past, and while these remain important industrial producers today, they would hardly be invoked as models for improvement of livelihoods and social welfare. A legitimate concern regarding many of the Asian NICs, including Thailand, may well be whether or not the international accumulation processes in which they are integrated will lead in the directions previously traversed by so many of the former 'miracle' economies of Latin America.

The argument of this book has been that the long-term historical processes that have led to the current conjuncture are crucial for understanding both contemporary successes and contemporary failures. More specifically, it has been argued that the role of the state in these stories of success and failure can be told more compellingly by seeing the state itself as a crucial site for the internationalization processes that have transformed the Thai political economy. There is no unified entity called Thailand that has been brought kicking and screaming into the era of 'globalization' by 'market forces'. Rather, internationalizing capitalist classes and class fractions have acted through the state (as well, of course, as other networks), transforming it as they do so, to create alliances that further the internationalization of capital. Internationalization of capital and internationalization of the state are thus interwoven processes extending (unevenly, like all capitalist development) across all of recent Thai history.

Neither the internationalization of the state nor the class processes it manifests provide a master key for unlocking the conundrums of development and industrial transformation. Many other aspects of development, including the gendering of proletarianization and the ideological transformations of elite intellectual culture, play integral roles. Emphasis on internationalization of the state is thus not designed to 'explain' various features of dependent development as if it were an independent causal variable. Instead, the effort has been to enrich previous analyses of Thai development by bringing to the fore a dimension of the process that has been inadequately interpreted under the hegemonic, 'territorially trapped' categories of the social sciences.

The internationalization processes highlighted in this analysis help identify more clearly the role of the state in Thailand's long-term process of industrial transformation. It has not been '*the* state', conceived as a territorially bounded entity or a discreet realm of activity, which contributes to economic growth and change by interacting with 'market forces' or 'international capital'. Rather, it has been geographically expansive classes and class fractions, acting in and through the state, which have transformed the political economy of Thailand, doing so simultaneously through economic and extra-economic means.

Thus, the internationalized Cold War state, which brought together US economic leaders and geopolitical strategists with Thai 'bureaucratic capitalists' and Sino-Thai entrepreneurs, fomented an industrial transformation that was predicated on capturing and squeezing the peasantry—and particularly on exploiting the labours of young, newly proletarianized women—while facilitating the growth of Bangkok-based, urban-industrial capital. The leading Thai capitalists of this period acted in and through the state, as well as through their international alliances. In doing so, they created a new 'hybrid' group of rural elites who facilitated intensification of primitive accumulation in upcountry regions such as the North. This in turn created the conditions for more intensive exploitation of labour, as much of the peasantry—deprived by military force of political powers that could improve its economic position—began to seek out the marginally higher wages available in urban-industrial employment.

The expanding and increasingly feminized proletariat was not simply allowed by the internationalized state to find its way into industrial occupations. State actors pushed and prodded the peasantry at one end (through agrarian policies that drew surplus from rural areas), while on the other end they directly facilitated industrial development and established mechanisms of industrial labour control that combined repression and co-optation. Moreover, while internationalization took the form of geographically expansive counter-insurgency/development projects, state actors prevented the formation of nationally cohesive political blocs that could resist or transform the elite agenda, utilizing Cold War rhetoric and military tools to destroy organizations such as the Peasant Federation of Thailand, the student movement, and the militant wing of the labour movement. This destruction of political opposition, whose effects linger to the present, heavily conditioned the terms on which peasants were proletarianized, making possible very rapid capital accumulation on the basis of extraordinarily high rates of exploitation.

As the Thai political economy has been further transformed under Japanese quasi-hegemony, this pattern of 'sweatshop'-based and internationalized industrial growth has remained at the core of development, though it has been overlaid by a veneer of higher value-added and more capital-intensive manufacturing activities, fomented largely by FDI and employing a very small fraction of the labour force. The Japanese presence in the Thai state has been less prominent than was the US presence, but this is in part because the prior process of internationalization within the Thai state already created the kind of orientation necessary for effective, internationalized accumulation involving Japanese investors. Moreover, the state continues to be a tool of class-based accumulation activities—even in the era of 'democratization'—as evidenced by the ascendancy of business elites within the state, including but not limited to the new generation of telecommunications tycoons.

By the mid-1980s, conditions became propitious for the internationalized state and the international class alliance of which it was a part to foment a more rapid expansion of industrial growth and transformation. As it did so, it used its geographical expansiveness to advantage, drawing on pre-existing international connections to spur a growth boom. This boom further exacerbated the socio-spatial unevenness of development in Thailand, but it also eventually allowed faster growth of manufacturing industry in certain upcountry regions of Thailand, such as the Chiang Mai–Lamphun area. The growth that occurred in these regions was, like the previous rounds of manufacturing growth in Bangkok, predicated heavily on marginalization of the peasantry, 'feminization' of manufacturing labour, and fragmentation of labour and political opposition—thus ensuring that growth on the periphery of the periphery would be of the 'sweatshop' variety and based heavily on international capital flows which stimulated relatively disarticulated growth.

The emphasis placed here on internationalization processes that have been active throughout modern Thai history contributes to the explanation of Thai industrialization and proletarianization, showing it to be an internationalized class project, with gendered dimensions, carried forward by economic and extra-economic means. Moreover, emphasis on international class alliances helps focus precisely how it is that this project has been successful on its own terms—fomenting rapid accumulation and high profit rates for internationalized elites—irrespective of its precise effects on different segments of rural society or the urban labour force. But the economic crisis in Thailand, and thus the capital accumulation project's moment of 'failure', can also be effectively interpreted through this geographical-historical materialist lens. It is part of the 'inner nature' of capitalism that its successes are the very condition of its subsequent failures, and Thailand's crisis manifests how such failures are enabled and conditioned by the same processes of internationalization that contributed to previous success.

The rise of the seemingly more nationalist Thai Rak Thai Party has so far failed to derail internationalizing tendencies within either the economy or the state, even if the ultimate outcomes of post-crisis restructuring are not yet determined. Rather than calling into question either internationalization or the interpretation of Thai development as based on it, restructuring and even the 'nationalist' backlash in Thailand seems to attest to the continuing if now more contested hegemony of internationalist forces. Thus, the economic crisis in Thailand differs from the preceding era of successful capitalist development only in that it has claimed victims among some of the previously privileged. For those who have been most marginalized in the process of successful development, neither the economic crisis nor its resolution are likely to resolve the problems of uneven development, though a protracted crisis could eventually create a context of intensified and internationalized social struggle—through which something better might be born.

BIBLIOGRAPHY

Agnew, J. A. (1994). 'The Territorial Trap: The Geographical Assumptions of International Relations Theory'. *Review of International Political Economy*, 1/1: 53–80.

——(1998). *Geopolitics: Re-Visioning World Politics*. London and New York: Routledge.

——and Corbridge, S. (1995). *Mastering Space: Hegemony, Territory and International Political Economy*. New York: Routledge.

Althusser, L. (1972). *Lenin and Philosophy*. New York: Monthly Review Press.

Amin, S. (1974). *Accumulation on a World Scale: A Critique of the Theory of Underdevelopment*. New York: Monthly Review Press.

Amnat Dangrua (1996). *Future Trends in the Establishment of Small-Scale Industry in Chiang Mai-Lamphun*. Chiang Mai: Ministry of Industry of Thailand.

Amsden, A. H. (1989). *Asia's Next Giant: South Korea and Late Industrialization*. New York: Oxford University Press.

——(1990). 'Third World Industrialization: "Global Fordism" or a New Model'. *New Left Review*, 182: 5–31.

——(1994). 'Why isn't the Whole World Experimenting with the East Asian Model to Develop?'. *World Development*, 22/4: 627–33.

——(1995). 'Like the Rest: South-East Asia's "Late" Industrialization'. *Journal of International Development*, 7/5: 791–9.

Anan Ganjanapan (1984). 'The Partial Commercialization of Rice Production in Northern Thailand, 1900–1980'. Ph.D. thesis, Cornell University, Ithaca, NY.

——(1989). 'Conflicts over the Deployment and Control of Labor in a Northern Thai Village', in G. P. Hart, A. Turton, and B. White (eds.), *Agrarian Transformations: Local Processes and the State in Southeast Asia*. Berkeley: University of California Press, 98–122.

Anchalee Singhanetra-Renard (1987). 'Non-Farm Employment and Female Labour Mobility in Northern Thailand', in J. H. Momsen and J. G. Townsend (eds.), *Geography of Gender in the Third World*. London: SUNY Press, 258–73.

Anderson, B. (1977). 'Withdrawal Symptoms: Social and Cultural Aspects of the October 6 Coup'. *Bulletin of Concerned Asian Scholars*, 9/3: 13–30.

——(1978). 'Studies of the Thai State: The State of Thai Studies', in E. B. Ayal (ed.), *The Study of Thailand: Analyses of Knowledge, Approaches, and Prospects in Anthropology, Art History, Economics, History, and Political Science*. Athens, Oh.: Ohio University Center for International Studies, 193–247.

——(1990). 'Murder and Progress in Modern Siam'. *New Left Review*, 181: 33–48.

——(1991). *Imagined Communities: Reflections on the Origin and Spread of Nationalism* (2nd edn.). London and New York: Verso.

Anek Laothamatas (1988). 'Business and Politics in Thailand: New Patterns of Influence'. *Asian Survey*, 28/4: 451–70.

——(1992*a*). *Business Associations and the New Political Economy of Thailand: From Bureaucratic Polity to Liberal Corporatism*. Boulder, Colo. and Singapore: Westview Press and Institute of Southeast Asian Studies.

Anek Laothamatus (1992*b*). 'The Politics of Structural Adjustment in Thailand: A Political Explanation of Economic Success', in A. J. MacIntyre and K. Jayasuriya (eds.), *The Dynamics of Policy Reform in South-east Asia and the South-west Pacific*. Singapore: Oxford University Press, 32–49.

——(1993). 'Sleeping Giant Awakens!: The Middle Class in Thai Politics'. *Asian Review*, 7: 78–125.

——(1994). 'From Clientelism to Partnership: Business–Government Relations in Thailand', in A. J. MacIntyre (ed.), *Business and Government in Industrializing Asia*. Ithaca, NY: Cornell University Press, 195–215.

Anuchat Poungsomlee and Ross, H. (1992). *Impacts of Modernisation and Urbanisation in Bangkok: An Integrative Ecological and Biosocial Study*. Salaya, Thailand: Institute for Population and Social Research.

Armstrong, W. R. and McGee, T. G. (1968). 'Revolutionary Change and the Third World City: A theory of Urban Involution'. *Civilisations*, 18/3: 353–77.

Atkinson, A. B. (1983). *The Economics of Inequality*. Oxford: Clarendon Press.

Ayal, E. B. (1992). 'Thailand's Development: The Role of Bangkok'. *Pacific Affairs*, 65/3: 353–67.

Baker, C. (2000). 'Thailand's Assembly of the Poor: Background, Drama, Reaction'. *South East Asia Research*, 8/1: 5–29.

——(2004). 'Pluto-Populism: Thaksin, Business, and Popular Politics', in P. Warr (ed.), *Thailand Beyond the Crisis*. London: Routledge.

Balassa, B. (ed.) (1981). *The Newly Industrializing Countries in the World Economy*. New York: Pergamon.

Bangkok Industrial Authority (BIA) (1994). Unpub. data on Bangkok factories. Bangkok: BIA.

Bank of Thailand (1965–2000). *Annual Report*. Bangkok: Bank of Thailand.

——(1976–97). Unpublished lending statistics. Bangkok: Bank of Thailand.

—— (1985–7). *Production, Investment, and Employment in Manufacturing, Trade and Construction Sectors*. Bankok: Bank of Thailand

——(1988–98). *Monthly Bulletin of Statistics*. Bangkok: Bank of Thailand.

——(1997). *Quarterly Bulletin*. Bangkok: Bank of Thailand (Dec.).

Becker, D. G. and Sklar, R. L. (1987). 'Why Postimperialism?', in D. Becker, J. Frieden, S. P. Schatz, and R. L. Sklar (eds.), *Postimperialism: International Capitalism and Development in the Late Twentieth Century*. Boulder, Colo. and London: Lynne Reinner Publishers, 1–18.

Beeson, M. (2001). 'Japan and Southeast Asia: The Lineaments of Quasi-Hegemony', in G. Rodan, K. Hewison, and R. Robison (eds.), *The Political Economy of South-East Asia: Conflicts, Crises, and Change* (2nd edn.). Melbourne: Oxford University Press, 283–306.

Bell, P. F. (1977). 'Marxist Theory, Class Struggle and the Crisis of Capitalism', in J. Schwartz (ed.), *The Subtle Anatomy of Capitalism*. Santa Monica, Calif.: Goodyear, 170–94.

——(1978). '"Cycles" of Class Struggle in Thailand'. *Journal of Contemporary Asia*, 8/1: 51–79.

——(1991). 'The Impact of the United States on the Development of the Social Sciences in Thailand', in J. A. Lent (ed.), *Studies in Third World Societies*, Social Science Models and Their Impact on the Third World no. 45. Williamsburg, Va.: Department of Anthropology, College of William and Mary, 95–116.

——(1992). 'Gender and Economic Development in Thailand', in P. Van Esterik and J. Van Esterik (eds.), *Gender and Development in Southeast Asia*. Montreal: Canadian Council for Southeast Asian Studies, 61–82.

——(1996). 'Development or Maldevelopment? The Contradictions of Thailand's Economic Growth', in M. Parnwell (ed.), *Uneven Development in Thailand*. Aldershot and Brookfield Vt.: Avebury, 49–62.

——(1997). 'Thailand's Economic Miracle: Built on the Backs of Women', in Virada Somswasdi and S. Theobald (eds.), *Women, Gender Relations and Development in Thai Society*, i. Chiang Mai: Women's Study Center, Chiang Mai University, 55–82.

Bello, W. F. (1994). *Dark Victory: The United States, Structural Adjustment, and Global Poverty*. London and Oakland, Calif.: Pluto Press and Institute for Food and Development Policy.

——(1998). 'East Asia: On the Eve of the Great Transformation?' *Review of International Political Economy*, 5/3: 424–44.

——and Rosenfeld, S. (1990). *Dragons in Distress: Asia's Miracle Economies in Crisis*. San Francisco: Institute for Food and Development Policy.

——Cunningham, S., and Poh, L. K. (1998). *A Siamese Tragedy: Development and Disintegration in Modern Thailand*. London, New York, Oakland, Calif. and Bangkok: Zed Books, Institute for Food and Development Policy, and White Lotus.

Benería, L. (1979). 'Reproduction, Production and the Sexual Division of Labour'. *Cambridge Journal of Economics*, 3/3: 203–25.

Bernard, M. (1999). 'East Asia's Tumbling Dominoes: Financial Crises and the Myth of the Regional Model', in L. Panitch and C. Leys (eds.), *Socialist Register 1999*. New York and London: Monthly Review Press, 178–208.

——and J. Ravenhill (1995). 'Beyond Product Cycles and Flying Geese: Regionalization, Hierarchy, and the Industrialization of East Asia'. *World Politics* 47/2: 171–209.

Bevacqua, R. (1998). 'Whither the Japanese Model? The Asian Economic Crisis and the Continuation of Cold War Politics in the Pacific Rim'. *Review of International Political Economy*, 5/3: 410–23.

Bhagwati, J. (1988). 'Export-promoting Trade Strategy: Issues and Evidence'. *World Bank Research Observer*, 3/1: 27–57.

Bhansoon Ladavalya and Vasant Siripool (1988). 'Summary of the Research Report on the Impact of the Regional City Development Project upon Outlying Areas: The Chiang Mai Case', Chiang Mai: Social Research Institute, Chiang Mai University.

Biersteker, T. J. (1995). 'The "Triumph" of Liberal Economic Ideas in the Developing World', in B. Stallings (ed.), *Global Change, Regional Response: The New International Context of Development*. Cambridge: Cambridge University Press, 174–96.

Blaut, J. M. (1987). *The National Question: Decolonizing the Theory of Nationalism*. London and Atlantic Highlands, NJ, Zed Books.

Block, F. (1977). 'The Ruling Class Does Not Rule: Notes on the Marxist Theory of the State'. *Socialist Review*, 7/3: 6–28.

Board of Investment of Thailand (BoI) (1993). *A Guide to Investing in Thailand (including a Guide to the Board of Investment)*. Bangkok: BoI.

——(1988–99). 'Investment Statistics'. Bangkok: BoI. (Unpub. quarterly data available in BoI library.)

Booth, A. (1999). 'Initial Conditions and Miraculous Growth: Why is South East Asia Different from Taiwan and South Korea?'. *World Development*, 27/2: 301–21.

Borden, W. S. (1984). *The Pacific Alliance: United States Foreign Economic Policy and Japanese Trade Recovery, 1947–1955*. Madison: University of Wisconsin Press.

Bowie, K. A. (1997). *Rituals of National Loyalty: An Anthropology of the State and the Village Scout Movement in Thailand.* New York: Columbia University Press.

Boyer, R. (1994). 'Do Labour Institutions Matter for Economic Development? A "Régulation" Approach for the OECD and Latin America with an Extension to Asia', in G. Rodgers (ed.), *Workers, Institutions and Economic Growth in Asia.* Geneva: International Institute for Labor Studies, 25–99.

Bradley, W. L. (1978). *Thailand, Domino by Default? The 1976 Coup and Implications for U.S. Policy, with an Epilogue on the October 1977 Coup*. Athens, Oh.: Ohio University Center for International Studies, Southeast Asia Program.

Brenner, N. (1999). 'Beyond State-Centrism? Space, Territoriality, and Geographical Scale in Globalization Studies'. *Theory and Society*, 28/1: 39–78.

Brenner, R. (1977). 'The Origins of Capitalist Development: A Critique of Neo-Smithian Marxism'. *New Left Review*, 104: 25–92.

——(1998). 'The Economics of Global Turbulence'. *New Left Review*, 229: 1–265.

Brown, A. and Frenkel, S. (1993). 'Union Unevenness and Insecurity in Thailand', in S. Frenkel (ed.), *Organized Labor in the Asia-Pacific Region: A Comparative Study of Trade Unionism in Nine Countries*. Ithaca, NY: ILR Press, 82–106.

——Bundit Thanachaisetavut, and Hewison, K. (2002). 'Labour Relations and Regulation in Thailand: Theory and Practice'. Working Paper no. 27, Southeast Asia Research Centre, City University of Hong Kong. Available at http://www.cityu.edu.hk/searc.

Brown, C. (1981). 'Mothers, Fathers, and Children: From Private to Public Patriarchy', in L. Sargent (ed.), *Women and Revolution: A Discussion of the Unhappy Marriage of Marxism and Feminism*. Boston: South End Press, 239–67.

Brown, I. (1978). 'British Financial Advisers in Siam in the Reign of King Chulalongkorn'. *Modern Asian Studies*, 12/2: 193–215.

——(1988). *The Elite and the Economy in Siam, c.1890–1920*. Singapore and New York: Oxford University Press.

——(1992) *The Creation of the Modern Ministry of Finance in Siam, 1885–1910*. London: Macmillan.

Bruneau, M. (1982). 'Mode of Production and Spatial Organization in Thailand: Process and Trends'. *Antipode*, 14/1: 1–10.

Bryan, R. (1987). 'The State and the Internationalisation of Capital: An Approach to Analysis'. *Journal of Contemporary Asia*, 17/3: 253–75.

Bullard, N., Bello, W., and Malhotra, K. (1998). 'Taming the Tigers: The IMF and the Asian Crisis', in K. S. Jomo (ed.), *Tigers in Trouble: Financial Governance, Liberalisation and Crises in East Asia.* Hong Kong: Hong Kong University Press, IPSR Books, University Press, White Lotus, and Zed Books, 85–136.

Bundit Thanachaisethavut (1998). *Trade Union Structure and Tripartite Systems in Thailand.* Bangkok: Arom Pongpangan Foundation and Friederich Ebert Stiftung.

Bureau of the Budget of Thailand (1995–7). *Thailand's Budget in Brief.* Bangkok: Bureau of the Budget.

Cardoso, F. H. and Faletto, E. (1979). *Dependency and Development in Latin America.* Berkeley: University of California Press.

Chai-anan Samudavanija (1995). 'Economic Development and Democracy', in M. Krongkaew (ed.), *Thailand's Industrialization and its Consequences*. London and New York: Macmillan and St. Martin's Press, 235–50.

——Kusuma Sanitwong Na Ayutthaya, and Suchit Bunbongkan (1990). *From Armed Suppression to Political Offensive*. Bangkok: Institute of Security and International Studies.

Chairat Charoensin-o-larn (1988). *Understanding Postwar Reformism in Thailand: A Reinterpretation of Rural Development*. Bangkok: Editions Duang Kamol.

Chaiyan Rajchagool (1994). *The Rise and Fall of the Thai Absolute Monarchy: Foundations of the Modern Thai State from Feudalism to Peripheral Capitalism*. Bangkok: White Lotus.

Chakrit Noranitipadungkarn and Hagensick, A. C. (1973). *Modernizing Chiengmai: A Study of Community Elites in Urban Development*. Bangkok: National Institute of Development Administration.

Chalongphob Sussangkarn (1993). 'Labour Markets', in P. G. Warr (ed.), *The Thai Economy in Transition*. Cambridge and New York: Cambridge University Press, 355–400.

Chandler and Thong-ek Law Offices (trans.) (1998). *Labour Protection Act B.E. 2541 (A.D. 1998)*. Bangkok: Nititham Publishing House.

Chayan Vaddhanaphuti (1993). 'Traditions of Village Study in Thailand', in P. Hirsch (ed.), *The Village in Perspective: Community and Locality in Rural Thailand*. Chiang Mai: Chiang Mai University Social Research Institute.

Chiang Mai Provincial Industry Office (PIO) (1992–7). Unpub. data on Chiang Mai industries. Chiang Mai: Chiang Mai PIO.

Chibber, V. (1999). 'Building a Developmental State: The Korean Case Reconsidered'. *Politics & Society*, 27/3: 309–46.

Chokchai Suttawet (1994). 'Industrial Relations Practices and Strategies: A Theoretical Construction for Thailand', Ph.D. thesis, Bielefeld University.

Chrobot, S. (1996). *Trade Unions in Transition: Present Situation and Structure of the Thai Labor Movement*. Bangkok: Friedrich Ebert Stiftung.

Claessens, S., Djankov, S., Lang, L. (1998). 'East Asian Corporates: Growth, Financing and Risks over the Last Decade', World Bank Working Paper, no. 2017, 27 Oct. Washington: World Bank.

Connors, M. K. (1999). 'Political Reform and the State in Thailand'. *Journal of Contemporary Asia*, 29/4: 202–26.

——(2003a). *Democracy and National Identity in Thailand*. London: RoutledgeCurzon.

——(2003b). 'The Reforming State: Security, Development and Culture in Democratic Times'. Working Paper no. 42, Southeast Asia Research Centre, City University of Hong Kong. Available at http://www.cityu.edu.hk/searc

Cox, R. W. (1981). 'Social Forces, States and World Orders: Beyond International Relations Theory'. *Millennium: Journal of International Studies*, 10/2: 126–55.

——(1987). *Production, Power, and World Order: Social Forces in the Making of History*. New York: Columbia University Press.

Crafts, N. (1999). 'East Asian Growth Before and After the Crisis'. IMF Staff Papers 46/2. Washington: International Monetary Fund.

Cravey, A. J. (1998). *Women and Work in Mexico's Maquiladoras*. Lanham, Md.: Rowman & Littlefield.

Crispin, S. (2001). 'Digitizing a Nation', *Far Eastern Economic Review*, 164/9 (8 Mar.), 30–2.

Cumings, B. (1981). *The Origins of the Korean War*, i. *Liberation and the Emergence of Separate Regimes, 1945–1947*. Princeton: Princeton University Press.

——(1984). 'The Origins and Development of the Northeast Asian Political Economy: Industrial Sectors, Product Cycles, and Political Consequences'. *International Organization*, 38/1: 1–40.

——(1990). *The Origins of the Korean War*, ii. *The Roaring of the Cataract, 1947–1950*. Princeton: Princeton University Press.

——(1999*a*). *Parallax Visions: Making Sense of American–East Asian Relations at the End of the Century*. Durham, NC: Duke University Press.

——(1999*b*). 'Webs with No Spiders, Spiders with No Webs: The Geneology of the Developmental State', in M. Woo-Cumings (ed.), *The Developmental State*. Ithaca, NY: Cornell University Press, 61–92.

Daranee Bunchorntavakul (1976). 'The Impact of the Growth Center on Rural Areas: A Case Study in the North of Thailand'. Rangsit, Thailand: Asian Institute of Technology.

Darling, F. C. (1965). *Thailand and the United States*. Washington: Public Affairs Press.

Davidson, P. and Ciambella, F. (1995). *Investment in Southeast Asia: Policy and Laws*. Singapore: Butterworth-Heinemann Asia, 233–89.

De Janvry, A. (1981). *The Agrarian Question and Reformism in Latin America*. Baltimore: Johns Hopkins University Press.

Deere, C. D. (1976). 'Rural Women's Subsistence Production in the Capitalist Periphery'. *Review of Radical Political Economics*, 8: 9–15.

Department of Labour Protection and Welfare of Thailand (1985–2000). *Yearbook of Labour Statistics*. Bangkok: Department of Labour Protection and Welfare.

Devillers, P. (1970). 'The Laotian Conflict in Perspective', in N. S. Adams and A. J. McCoy (eds.), *Laos: War and Revolution*. New York: Harper & Row, 37–52.

Devine, J. (1994). 'The Causes of the 1929–33 Great Collapse: A Marxian Interpretation'. *Research in Political Economy*, 14: 119–94.

Deyo, F. C. (1989). *Beneath the Miracle: Labor Subordination in the New Asian Industrialism*. Berkeley: University of California Press.

——(1995*a*). 'Capital, Labor, and State in Thai Industrial Restructuring: The Impact of Global Economic Transformations', in J. Böröcz and D. A. Smith (eds.), *A New World Order? Global Transformations in the Late Twentieth Century*. Westport, Conn.: Greenwood Press, 131–44.

——(1995*b*). 'Human Resources Strategies and Industrial Restructuring in Thailand', in S. Frenkel and J. Harrod (eds.), *Industrialization and Labor Relations: Contemporary Research in Seven Countries*. Ithaca, NY: ILR Press, 23–36.

Dicken, P. (1997). 'Transnational Corporations and Nation-states'. *International Social Science Journal*, 151: 77–99.

——and Yeung, H. W.-c. (1999). 'Investing in the Future: East and Southeast Asian Firms in the Global Economy', in K. Olds, P. Dicken, P. F. Kelly, L. Kong, and H. W.-c. Yeung (eds.), *Globalization and the Asia-Pacific: Contested Territories*. London: Routledge, 107–28.

Dixon, C. J. (1999). *The Thai Economy: Uneven Development and Internationalisation*. London and New York: Routledge.

——and Parnwell, M. J. G. (1991). 'Thailand: The Legacy of Non-Colonial Development in South-East Asia', in C. Dixon and M. J. Heffernan (eds.), *Colonialism and Development in the Contemporary World*. London: Mansell, 204–25.

Doner, R. F. (1991). *Driving a Bargain: Automobile Industrialization and Japanese Firms in Southeast Asia*. Berkeley: University of California Press.

—— and A. Ramsey (1994). 'Thailand in the Pacific Rim Garment Industry', in E. Bonacich, L. Cheng, N. Chinchilla, N. Hamilton, and P. Ong (eds.), *Global Production: The Apparel Industry in the Pacific Rim*. Philadelphia: Temple University Press, 180–96.

—— —— (1997). 'Competitive Clientelism and Economic Governance: The Case of Thailand', in S. Maxfield and B. R. Schneider (eds.), *Business and the State in Developing Countries*. Ithaca, NY: Cornell University Press, 237–76.

Donner, W. (1978). *The Five Faces of Thailand: An Economic Geography*. New York: St. Martin's Press.

Douglass, M. (1984). *Regional Integration of the Capitalist Periphery: The Central Plans of Thailand*. The Hague: Institute of Social Studies.

Duménil, G. and Lévy, D. (1993). *The Economics of the Profit Rate: Competition, Crises, and Historical Tendencies in Capitalism*. Aldershot and Brookfield, Vt.: Edward Elgar.

Duncan, S. (1994). 'Theorising Differences in Patriarchy'. *Environment and Planning A*, 26/8: 1177–94.

Economist Intelligence Unit (EIU) (1992). *Country Report. Thailand, Myanmar (Burma), 1992–93*. London: EIU.

Escobar, A. (1995). *Encountering Development: The Making and Unmaking of the Third World*. Princeton: Princeton University Press.

Evans, P. (1979). *Dependent Development: The Alliance of Multinational, State, and Local Capital in Brazil*. Princeton: Princeton University Press.

——(1987). 'Class, State, and Dependence in East Asia: Lessons for Latin Americanists', in F. C. Deyo (ed.), *The New Asian Industrialism*. Ithaca, NY: Cornell University Press, 203–24.

——(1989). 'Predatory, Developmental, and Other Apparatuses: A Comparative Political Economy Perspective on the Third World State'. *Sociological Forum*, 4/4: 561–87.

——(1995). *Embedded Autonomy: States and Industrial Transformation*. Princeton: Princeton University Press.

Evers, H.-D. (1987). 'Trade and State Formation: Siam in the Early Bangkok Period'. *Modern Asian Studies*, 21/4: 751–71.

Fairclough, G. (1997). 'Feeling Squeezed'. *Far Eastern Economic Review*, 160/2 (9 Jan.), 81.

Falkus, M. (1995). 'Thai Industrialization: An Overview', in M. Krongkaew (ed.), *Thailand's Industrialization and its Consequences*. New York: St. Martin's Press, 13–32.

Faust, D. and Nagar, R. (2001). 'English Medium Education, Social Fracturing, and the Politics of Development in Postcolonial India'. *Economic and Political Weekly*, 37/8 (28 July–3 Aug.), 2878–83.

Feeny, D. (1982). *The Political Economy of Productivity: Thai Agricultural Development, 1880–1975*. Vancouver: University of British Columbia Press.

Ferguson, T. (1984). 'From Normalcy to New Deal: Industrial Structure, Party Competition, and American Public Policy in the Great Depression'. *International Organization*, 38/1: 41–94.

Fineman, D. (1997). *A Special Relationship: The United States and Military Government in Thailand, 1947–1958*. Honolulu: University of Hawai'i Press.

Fisher, C. A. (1964). *South-East Asia: A Social, Economic, and Political Geography*. London and New York: Methuen and Dutton.

Flatters, F. and Mingsarn Santikarn Kaosa-ard (1994). 'Global and Regional Economic Change: Implications for Northern Thailand'. *TDRI Quarterly Review*, 9/1: 8–15.

Flood, E. T. (1977). 'The United States and the Military Coup in Thailand: A Background Study', Berkeley, Calif.: Indochina Resource Center.

Forsyth, T. (1994). 'Shut Up or Shut Down: How a Thai Medical Agency was Closed After it Questioned Worker Safety at a Factory Owned by Thailand's Largest Employer'. *Asia Inc.* (Apr.), 30–7.

Foucault, M. (1977). *Discipline and Punish: The Birth of the Prison*. New York: Pantheon Books.

Frank, A. G. (1966). 'The Development of Underdevelopment'. *Monthly Review*, 18/4: 17–31.

Fraser, T. M. (1966). *Fishermen of South Thailand: The Malay Villagers*. New York: Holt, Rinehart and Winston.

Freeland, R. M. (1985). *The Truman Doctrine and the Origins of McCarthyism: Foreign Policy, Domestic Politics, and Internal Security, 1946–1948*. New York: New York University Press.

Frenkel, S. (1993). 'Variations in Patterns of Trade Unionism: A Synthesis', in S. Frenkel (ed.), *Organized Labor in the Asia-Pacific Region: A Comparative Study of Trade Unionism in Nine Countries*. Ithaca, NY: ILR Press, 309–46.

Galtung, J. (1971). 'A Structural Theory of Imperialism'. *Journal of Peace Studies*, 8/2: 81–117.

Gawin Chutima (1990). *The Rise and Fall of the Communist Party of Thailand (1973–1987)*. Canterbury: Centre of South-East Asian Studies.

Gilbert, A. and Gugler, J. (1992). *Cities, Poverty, and Development: Urbanization in the Third World*. Oxford: Oxford University Press.

Girling, J. L. S. (1981). *Thailand: Society and Politics*. Ithaca, NY: Cornell University Press.

Glassman, J. (2001). 'Women Workers and the Regulation of Health and Safety on the Industrial Periphery: The Case of Northern Thailand', in I. Dyck, N. Lewis, and S. McLafferty (eds.), *Geographies of Women's Health*. London and New York: Routledge, 61–87.

——(2002). 'From Seattle (and Ubon) to Bangkok: The Scales of Resistance to Corporate Globalization'. *Environment and Planning D: Society and Space*, 20/5: 513–33.

——(2003). 'The Spaces of Economic Crisis: Asia and the Reconfiguration of Neo-Marxist Crisis Theory'. *Studies in Comparative International Development*, 37/4: 31–63.

——and Carmody, P. (2001). 'Structural Adjustment in East and Southeast Asia: Lessons from Latin America'. *Geoforum*, 32/1: 77–90.

——and Samatar, A. I. (1997). 'Development Geography and the Third-World State'. *Progress in Human Geography*, 21/2: 164–98.

Glyn, A. (1997). 'Does Aggregate Profitability Really Matter?'. *Cambridge Journal of Economics*, 21/5: 593–619.

Gold, T. B. (1986). *State and Society in the Taiwan Miracle*. Armonk, NY: M. E. Sharpe.

Gray, J. (1990). 'The Road to the City: Young Women and Transition in Northern Thailand', Ph.D. thesis, Macquarie University, Sydney.

Grit Permtanjit (1982). *Political Economy of Dependent Capitalist Development: Study on the Limits of the Capacity of the State to Rationalize in Thailand*. Bangkok: Chulalongkorn University Social Research Institute.

Halm, G. E. (1973). 'The Department of Labor in Thailand', Report by the Labor Department Advisor, U.S. Department of Labor and United States Operations Mission to Thailand, Agency for International Development. Rosslyn, Va., USAID Library.

Hart-Landsberg, M. (1993). *The Rush to Development: Economic Change and Political Struggle in South Korea*. New York: Monthly Review Press.

——and Burkett, P. (1998). 'Contradictions of Capitalist Industrialization in East Asia: A Critique of "Flying Geese" Theories of Development'. *Economic Geography*, 74/2: 87–110.

—— ——(2001). 'Economic Crisis and Restructuring in South Korea: Beyond the Free Market-Statist Debate'. *Critical Asian Studies*, 33/3: 403–30.

Harvey, D. (1982). *The Limits to Capital*. Oxford: Blackwell.

——(1985). *The Urbanization of Capital*. Oxford: Blackwell.

——(1989). *The Condition of Postmodernity: An Enquiry into the Origins of Cultural Change*. Oxford: Blackwell.

Hatch, W. and Yamamura, K. (1996). *Asia in Japan's Embrace: Building a Regional Production Alliance*. Cambridge and New York: Cambridge University Press.

Herod, A. (2001). *Labor Geographies: Workers and the Landscapes of Capitalism*. New York and London: Guilford Press.

——Ó Tuathail, G., and Roberts, S. M. (eds.) (1998). *An Unruly World? Globalization, Governance and Geography*. London and New York: Routledge.

Hersh, J. (1997). 'The Impact of US Strategy: Making Southeast Asia Safe for Capitalism', in J. D. Schmidt, J. Hersh, and N. Fold (eds.), *Social Change in Southeast Asia*. Harlow, Essex: Longman, 23–39.

Hewison, K. (1989*a*). *Bankers and Bureaucrats: Capital and the Role of the State in Thailand*. New Haven: Yale University Southeast Asia Studies, Yale Center for International and Area Studies.

——(1989*b*). *Power and Politics in Thailand: Essays in Political Economy*. Manila: Journal of Contemporary Asia Publishers.

——(1997). 'Introduction: Power, Oppositions and Democratisation', in K. Hewison (ed.), *Political Change in Thailand: Democracy and Participation*. London and New York: Routledge, 1–20.

——(2000). 'Thailand's Capitalism Before and After the Economic Crisis', in R. Robison, M. Beeson, K. Jayasuriya and H.-R. Kim (eds.), *Politics and Markets in the Wake of the Asian Crisis*. London and New York: Routledge, 192–211.

——(2003). 'The Politics of Neo-Liberalism: Class and Capitalism in Contemporary Thailand', Working Paper no. 45, Southeast Asia Research Centre, City University of Hong Kong. Available at http://www.cityu.edu.hk/searc.

Hewison, K. and Brown, A. (1994). 'Labour and Unions in an Industrialising Thailand'. *Journal of Contemporary Asia*, 24/4: 483–514.

——and Rodan, G. (1994). 'Whatever Happened to the Revolution? The Decline of the Left in Southeast Asia', in *Socialist Register 1994*. London: Merlin Press: 235–62.

Hirsch, P. (1990). *Development Dilemmas in Rural Thailand*. Singapore and New York: Oxford University Press.

Hirst, P. Q. and Thompson, G. (1999). *Globalization in Question: The International Economy and the Possibilities of Governance*. Cambridge and Malden, Mass.: Polity and Blackwell.

Hogan, M. J. (1987). *The Marshall Plan: America, Britain, and the Reconstruction of Western Europe, 1947–1952*. Cambridge and New York: Cambridge University Press.

Holm, D. F. (1977). 'The Role of the State Railways in Thai History, 1892–1932', Ph.D. thesis, Yale University.

Hook, G. D. (2001). 'Japan's Role in the East Asian Political Economy: An Emerging Region?', in G. D. Hook and H. Harukiyo (eds.), *The Political Economy of Japanese Globalization*. London and New York: Routledge: 40–55.

Hsing, Y.-t. (1998). *Making Capitalism in China: The Taiwan Connection*. New York: Oxford University Press.

Hussey, A. (1993). 'Rapid Industrialization in Thailand'. *Geographical Review*, 83/1: 14–28.

Ichikawa, N. (1991). 'Geographic Implications of Foreign Investment in Thailand's Industrialization'. *Southeast Asian Journal of Social Science*, 14/1–2: 64–81.

Industrial Estate Authority of Thailand (IEAT) (1995–6). 'Factories in Industrial Estates, 1995/1996'. Bangkok: Industrial Estate Authority of Thailand.

Industrial Finance Corporation of Thailand (IFCT) (1960–97). *Annual Report*. Bangkok: Industrial Finance Corporation of Thailand.

International Cooperation Administration (ICA). (1956). 'Summary of the Labor Situation in Thailand'. Washington: US Department of Labor.

——(1961). 'The Development of Thailand: Report of a Survey Team', Howard Rothman Bowen Collection, University of Iowa Library.

International Labor Organization (ILO) (1965–72). *Yearbook of Labor Statistics*. Geneva: International Labor Organization.

International Monetary Fund (IMF) (1997). 'Thailand Letter of Intent, 14 August 1997'. Available at http://www.imf.org/external/country/THA/index.htm.

——(1998). 'Thailand Letter of Intent, 1 December 1998'. Available at http://www.imf.org/external/country/THA/index.htm.

Jamaree Pitackwong (1996). 'Disorganized Development: Changing Forms of Work and Livelihood in Rural Northern Thailand', Ph.D. thesis, School of Oriental and African Studies, University of London.

Jansen, K. (1991). 'Thailand: The Next NIC?'. *Journal of Contemporary Asia*, 21/1: 13–30.

——(1997). *External Finance in Thailand's Development: An Interpretation of Thailand's Growth Boom*. Basingstoke and New York: Macmillan and St. Martin's Press.

Jenkins, R. (1987). *Transnational Corporations and Uneven Development: The Internationalization of Capital and the Third World*. London and New York: Methuen.

——(1991). 'The Political Economy of Industrialization: A Comparison of Latin American and East Asian Newly Industrializing Countries'. *Development and Change*, 22/2: 197–231.

Jessop, B. (1990*a*). 'Regulation Theories in Retrospect and Prospect'. *Economy and Society*, 19/2: 153–216.

——(1990*b*). *State Theory: Putting the Capitalist State in its Place*. Cambridge: Polity Press.

Ji Giles Ungpakorn (1997). *The Struggle for Democracy and Social Justice in Thailand*. Bangkok: Arom Pongpangan Foundation.

Johnson, C. (1982). *MITI and the Japanese Miracle: The Growth of Industrial Policy, 1925–1975*. Stanford, Calif.: Stanford University Press.

——(2000). *Blowback: The Costs and Consequences of American Empire*. New York: Metropolitan Books.

Johnston, D. B. (1975). 'Rural Society and the Rice Economy in Thailand', Ph.D. thesis, Yale University.

Jomo, K. S. (1997). *Southeast Asia's Misunderstood Miracle: Industrial Policy and Economic Development in Thailand, Malaysia and Indonesia*. Boulder, Colo.: Westview Press.

——(ed.) (1998). *Tigers in Trouble: Financial Governance, Liberalisation and Crises in East Asia*. Hong Kong: Hong Kong University Press, IPSR Books, University Press, White Lotus, and Zed Books.

Jones, A. (1998). 'Re-theorising the Core: A "Globalized" Business Elite in Santiago, Chile'. *Political Geography*, 17/3: 295–318.

Kahin, G. M. (1986). *Intervention: How America Became Involved in Vietnam*. New York: Knopf.

Kakwani, N. and Medhi Krongkaew (1996). 'Big Reduction in "Poverty"'. *Bangkok Post, Economic Review Year-End 1996*: 21–3.

Kanchada Poonpanich (1988). 'The Beginning of the Labor Movement in Thailand (1900–1930)'. *Asian Review*, 2: 24–53.

Kasian Tejapira (1993). 'Independent Siam and Colonial Burma: A Comparative Historical Perspective'. *Asian Review*, 7: 1–62.

——(2002). 'Post-Crisis Economic Impasse and Political Recovery in Thailand: The Resurgence of Economic Nationalism'. *Critical Asian Studies* 34/3: 323–56.

Kato, T. and Steven, R. (1991). 'Is Japanese Capitalism Post-Fordist? Papers of the Japanese Studies Centre, no. 16. Melbourne: Monash University.

Kenney, M. and Florida, R. (1988). 'Beyond Mass Production: Production and the Labor Process in Japan'. *Politics and Society*, 16/1: 121–58.

—— ——(1989). 'Japan's Role in a Post-Fordist Age'. *Futures*, 21/2: 136–51.

Kholi, A. (1999). 'Where do High-Growth Political Economies Come From? The Japanese Lineage of Korea's "Developmental State"', in M. Woo-Cumings (ed.), *The Developmental State*. Ithaca, NY: Cornell University Press, 93–136.

Klare, M. T. and Kornbluh, P. (1988). *Low Intensity Warfare: Counterinsurgency, Proinsurgency, and Antiterrorism in the Eighties*. New York: Pantheon Books.

Kolko, G. (1985). *Anatomy of a War: Vietnam, the United States, and the Modern Historical Experience*. New York: Pantheon Books.

——(1988). *Confronting the Third World: United States Foreign Policy, 1945–1980*. New York: Pantheon Books.

Kolko, G. (1997). *Vietnam: Anatomy of a Peace*. London and New York: Routledge.

Kolko, J. and Kolko, G. (1972). *The Limits of Power: The World and United States Foreign Policy, 1945–1954*. New York: Harper & Row.

Krugman, P. (1994). 'The Myth of Asia's Miracle'. *Foreign Affairs*, 73/6: 62–78.

Kumpol Puapanichya and Panayotou, T. (1985). 'Output Supply and Input Demand in Rice and Upland Crop Production: The Quest for Higher Yields in Thailand', in T. Panayotou (ed.), *Food Policy Analysis in Thailand*. Bangkok: Agricultural Development Council, 23–99.

Kunio, Y. (1988). *The Rise of Ersatz Capitalism in South-East Asia*. Singapore and New York: Oxford University Press.

Lamphun Provincial Industry Office (PIO) (1992–7). Unpub. data on Lamphun industries. Lamphun: Lamphun PIO.

Lansdale, E. G. (1972). *In the Midst of Wars: An American's Mission to Southeast Asia*. New York: Harper & Row.

Lauridsen, L. (1998). 'Thailand: Causes, Conduct, Consequences', in K. S. Jomo (ed.), *Tigers in Trouble: Financial Governance, Liberalisation and Crises in East Asia*. Hong Kong: IPSR Books, University Press, White Lotus, Zed Books, and Hong Kong University Press, 137–61.

——(1999). 'Policies and Institutions of Industrial Deepening and Upgrading in Thailand I: The Basic Industry Strategy in Upstream Petrochemicals'. Working Paper no. 14, International Development Studies, Roskilde University.

Lazonick, W. (1995). 'Cooperative Employment Relations and Japanese Economic Growth', in J. Schor and J.-I. You (eds.), *Capital, the State, and Labour: A Global Perspective*. Brookfield, Vt.: Edward Elgar, 70–110.

Leffler, M. P. (1992). *A Preponderance of Power: National Security, the Truman Administration, and the Cold War*. Stanford, Calif.: Stanford University Press.

Leftwich, A. (1995). 'Bringing Politics Back In: Towards a Model of the Developmental State'. *Journal of Development Studies*, 31/3: 400–27.

Lim, L. L. (1993). 'The Feminization of Labor in the Asia-Pacific Rim Countries: From Contributions to Economic Dynamism to Bearing the Brunt of Structural Adjustments', in N. Ogawa, G. W. Jones, and J. G. Williamson (eds.), *Human Resources in Development Along the Asia-Pacific Rim*. Singapore: Oxford University Press, 175–209.

Lim, T. C. (1998). 'Power, Capitalism, and the Authoritarian State in South Korea'. *Journal of Contemporary Asia*, 28/4: 457–83.

Limqueco, P., McFarlane, B. J., and Odhnoff, J. (1989). *Labour and Industry in ASEAN*. Manila: Journal of Contemporary Asia Publishers.

Little, I. M. D. (1981). 'The Experience and Causes of Rapid Labour-Intensive Development in Korea, Taiwan Province, Hong Kong and Singapore, and the Possibilities of Emulation', in E. Lee (ed.), *Export-Led Industrialization and Development*. Geneva: International Labor Organization, Asian Employment Programme.

Lobe, T. (1977). *United States National Security Policy and Aid to the Thailand Police*. Denver, Colo.: University of Denver Graduate School of International Studies.

London, B. (1980). *Metropolis and Nation in Thailand: The Political Economy of Uneven Development*. Boulder, Colo.: Westview Press.

——(1985). 'Thai City-Hinterland Relationships in an International Context: Development as Social Control in Northern Thailand', in M. Timberlake (ed.), *Urbanization in the World Economy*. Orlando, Fla.: Academic Press: 207–30.

Mabry, B. D. (1979). *The Development of Labor Institutions in Thailand*. Ithaca, NY: Southeast Asia Program, Cornell University.

McCargo, D. (1997). *Chamlong Srimuang and the New Thai Politics*. New York: St. Martin's Press.

McClintock, M. (1992). *Instruments of Statecraft: US Guerrilla Warfare, Counterinsurgency, and Counter-Terrorism, 1940–1990*. New York: Pantheon Books.

McCoy, A. W. (1991). *The Politics of Heroin: CIA Complicity in the Global Drug Trade*. Brooklyn, NY: Lawrence Hill Books.

McGehee, R. W. (1983). *Deadly Deceits: My 25 Years in the CIA*. New York: Sheridan Square Publications.

McMichael, P. and Myhre, D. (1990). 'Global Regulation vs. the Nation-State: Agro-Food Systems and the New Politics of Capital'. *Review of Radical Political Economics*, 22/1: 59–77.

Mallet, M. (1978). 'Causes and Consequences of the October '76 Coup'. *Journal of Contemporary Asia*, 8/1: 80–103.

Manat Suwan, Udom Kerdphibul, and Chukiat Leesuwan (1992). 'Impacts of Industrialization Upon the Village's Life in Northern Thailand'. Unpub. MS. Chiang Mai: Faculty of Social Science, Chiang Mai University.

Markusen, A. R. (1999). 'National Contexts and the Emergence of Second Tier Cities', in A. R. Markusen, Y.-S. Lee, and S. DiGiovanna (eds.), *Second Tier Cities: Rapid Growth Beyond the Metropolis*. Minneapolis: University of Minnesota Press, 65–94.

Marx, K. (1967). *Capital*, i. New York: International Publishers.

Massey, D. B. (1995). *Spatial Divisions of Labour: Social Structures and the Geography of Production*. London: Macmillan.

Mathana Phananiramai and Pawadee Tongudai (1994). *Final Report on Labor Welfare in Thailand*. Bangkok: Thailand Development Research Institute.

Mazumdar, D. (1994). 'Labor Markets and Adjustments in Open Asian Economies: The Republic of Korea and Malaysia'. *World Bank Economic Review*, 7/3: 349–80.

Medhi Krongkaew (1993). 'Poverty and Income Distribution', in P. G. Warr (ed.), *The Thai Economy in Transition*. Cambridge and New York: Cambridge University Press, 401–37.

——(1995). 'Contributions of Agriculture to Industrialization', in Medhi Krongkaew (ed.), *Thailand's Industrialization and its Consequences*. London and New York: Macmillan and St. Martin's Press: 33–65.

——(1996). 'Thailand: Poverty Assessment Update'. Bangkok: Economic Research and Training Center, Faculty of Economics, Thammasat University.

——Pranee Tinakorn, and Suphat Suphachalasai. (1992). 'Rural Poverty in Thailand: Policy Issues and Responses'. *Asian Development Review*, 10/1: 199–225.

Mills, M. B. (1997). 'Contesting the Margins of Modernity: Women, Migration, and Consumption in Thailand'. *American Ethnologist*, 24/1: 37–61.

——(1999). *Thai Women in the Global Labor Force: Consuming Desires, Contested Selves*. New Brunswick, USA: Rutgers University Press.

Mingsarn Kaosa-ard (2003). 'Poverty and Globalisation', in Mingsarn Kaosa-ard and J. Dore (eds.), *Social Challenges for the Mekong Region*. Chiang Mai: Chiang Mai University Social Research Institute: 81–108.

Mingsarn Kaosa-ard, Felt, E. O., Azumi, E. Y., Pednekar, S. S., Pornpen Wijukprasert, and Thippawan Keawmaesri. (1995). *Women in Development: Enhancing the Status of Rural Women in Northern Thailand.* Bangkok: Thailand Development Research Institute.

Morell, D. and Chai-anan Samudivanija (1981). *Political Conflict in Thailand: Reform, Reaction, Revolution.* Cambridge, Mass.: Oelgeschlager, Gunn and Hain.

Muecke, M. A. (1980). 'The Village Scouts of Thailand'. *Asian Survey*, 20/4: 407–27.

Murray, R. (1975). 'The Internationalization of Capital and the Nation State', in H. K. Radice (ed.), *International Firms and Modern Imperialism: Selected Readings.* Harmondsworth and Baltimore: Penguin Books: 107–34.

Muscat, R. J. (1990). *Thailand and the United States: Development, Security, and Foreign Aid.* New York: Columbia University Press.

——(1994). *The Fifth Tiger: A Study of Thai Development Policy.* Helsinki and Armonk, NY: United Nations University Press and M. E. Sharpe.

Napaporn Ativanichayapong (2002). 'Social Movement Unionism and Economic Unionism in Thailand'. *Asian Review* 15: 79–105.

Narathip Aramwattanano (1991). 'Factors Encouraging Investment by Foreign and Thai Companies in the Eastern Seaboard Region of Thailand', DBA thesis, US International University.

Nash, J. C. and Fernández-Kelly, M. P. (1983). *Women, Men, and the International Division of Labor.* Albany, NY: State University of New York Press.

National Economic and Social Development Board of Thailand (NESDB) (1970–96). *Capital Stock of Thailand.* Available online at http://www.nesdb.go.th/data_index/macroeconomic.html.

——(1970–2000). *National Accounts.* Bangkok: NESDB.

——(1981–2000). *Gross Regional Product (GRP) Statistics.* Bangkok: NESDB.

National Statistical Office of Thailand (NSO) (1978). *Agricultural Census Report.* Bangkok: NSO.

——(1997). *Industrial Census.* Bangkok: NSO.

——(1986–96). *Household Socio-Economic Survey (SES).* Bangkok: NSO.

——(1989–94). *Industrial Survey.* Bangkok: NSO.

——(1983–2000). *Labour Force Survey (LFS)*, February Round. Bangkok: NSO.

——(1960–2000). *Population and Housing Census.* Bangkok: NSO.

——(1978–1995). *Statistical Yearbook.* Bangkok: NSO.

Neher, A. (1980). 'Prelude to Alliance: The Expansion of the American Economic Role in Thailand during the 1940s', Ph.D. thesis, Northern Illinois University.

Nelson, K. (1986). 'Labor Demand, Labor Supply and the Suburbanization of Low-Wage Office Work', in A. J. Scott and M. Storper (eds.), *Production, Work, Territory: The Geographical Anatomy of Industrial Capitalism.* Boston: Allen & Unwin: 149–71.

Nikom Chandravithun (1982). *Thai Labour, A Long Journey.* Bangkok: Thai Watana Panich.

——(1995). 'Thailand: The Social Costs of Becoming the Fifth Tiger', The Woodrow Wilson Center Asia Program Occasional Paper, no. 68.

——and Vause, W. G. (1994). *Thailand's Labor and Employment Law: A Practical Guide.* Bangkok: Manager Publishing.

Nipon Poapongsakorn and Prayong Netayarak (1989). 'Regional Variations in Rural Interest Rates'. Unpub. MS. Bangkok: Faculty of Economics, Thammasat University.

Northern Region Industrial Estate (NRIE) (1997). *Northern Region Industrial Estate*. Lamphun, Thailand: Northern Region Industrial Estate. (Brochure, in Thai.)

——(1998). *Northern Region Industrial Estate. The Industrial Estate Authority of Thailand*. Lamphun, Thailand: Northern Region Industrial Estate. (Brochure, in Thai.)

Notermans, T. (1997). 'Social Democracy and External Constraints', in R. Cox (ed.), *The New Realism: Perspectives on Multilateralism and World Order*. Tokyo and New York: United Nations Press and St. Martin's Press, 201–39.

Ockey, J. (1992). 'Business Leaders, Gangsters, and the Middle Class', Ph.D. thesis, Cornell University.

O'Connor, J. (1973). *The Fiscal Crisis of the State*. New York: St. Martin's Press.

Ogle, G. E. (1990). *South Korea: Dissent Within the Miracle*. London: Zed Books.

Ohmae, K. (1995). *The End of the Nation State: The Rise of Regional Economies*. New York: Free Press.

ORSTOM (French Scientific Research Institute for Development by Cooperation) (1997). Unpub. data from Northern Thailand industry survey.

Overbeek, H., and van der Pijl, K. (1993). 'Restructuring Capital and Restructuring Hegemony: Neo-Liberalism and the Unmaking of the Post-War Order', in H. Overbeek (ed.), *Restructuring Hegemony in the Global Political Economy: The Rise of Transnational Neo-Liberalism in the 1980s*. London and New York: Routledge, 1–27.

Palloix, C. (1975). 'The Internationalization of Capital and the Circuit of Social Capital', in H. K. Radice (ed.), *International Firms and Modern Imperialism: Selected Readings*. Harmondsworth: Penguin Books: 63–88.

Panitch, L. (1994). 'Globalisation and the State', in R. Miliband and L. Paniteh (eds.), *Socialist Register 1994*. London: Merlin Press: 60–93.

Parnwell, M. J. G. (1992). 'Confronting Uneven Development in Thailand: The Potential Role of Rural Industries'. *Malaysian Journal of Tropical Geography*, 22/1: 51–62.

——(ed.) (1996). *Uneven Development in Thailand*. Aldershot and Brookfield Vt.: Avebury.

Parsons, T. (1966). *Societies: Evolutionary and Comparative Perspectives*. Englewood Cliffs, NJ: Prentice-Hall.

Pastor, M. (1987). *The International Monetary Fund and Latin America: Economic Stabilization and Class Conflict*. Boulder, Colo.: Westview Press.

Pasuk Phongpaichit (1982). *From Peasant Girls to Bangkok Masseuses*. Geneva: International Labor Office.

——and Baker, C. J. (1995). *Thailand: Economy and Politics* (1st edn.). Oxford and New York: Oxford University Press.

—— ——(1997). *Thailand's Boom!* Chiang Mai: Silkworm Books.

—— ——(1998). *Thailand's Boom and Bust*. Chiang Mai: Silkworm Books.

—— ——(2000). *Thailand's Crisis*. Chiang Mai: Silkworm Books.

—— ——(2002). '"The Only Good Populist is a Rich Populist": Thaksin Shinawatra and Thailand's Democracy'. Working Paper no. 36, Southeast Asia Research Centre, City University of Hong Kong. Available at http://www.cityu.edu.hk/searc.

—— ——(forthcoming). 'Pluto-Populism in Thailand: Business Remaking Politics', in J. Sidel and E.-L. Hedmann (eds.), *Populism and Reformism in Southeast Asia: The Threat and Promise of New Politics*. New Haven: Yale University Press.

Pasuk Phongpaichit and Sungsidh Piriyarangsan (1994). *Corruption and Democracy in Thailand.* Bangkok: Political Economy Centre, Faculty of Economics, Chulalongkorn University.

—— —— and Nualnoi Treerat. (1996). *Challenging Social Exclusion: Rights and Livelihood in Thailand.* Geneva: International Institute for Labor Studies and United Nations Development Programme.

—— —— ——(1998). *Guns, Girls, Gambling, Ganja: Thailand's Illegal Economy and Public Policy.* Chiang Mai: Silkworm Books.

Paterson, T. G. (ed.) (1989). *Kennedy's Quest for Victory: American Foreign Policy, 1961–1963.* New York and Oxord: Oxford University Press.

Peck, J. and Miyamachi, Y. (1994). 'Regulating Japan? Regulation Theory Versus the Japanese Experience'. *Environment and Planning D*, 12/6: 639–74.

Pempel, T. J. (1999). 'The Developmental Regime in a Changing World Economy', in M. Woo-Cumings (ed.), *The Developmental State.* Ithaca, NY: Cornell University Press, 137–81.

Pendakur, V. S. (1993). 'Congestion Management and Air Quality: Lessons from Bangkok and Mexico City'. *Asian Journal of Environmental Management*, 1/2: 53–65.

Petras, J. F. (1978). 'Aspects of Class Formation in the Periphery', in J. Petras (ed.), *Critical Perspectives on Imperialism and Social Class in the Third World.* New York: Monthly Review Press: 63–83.

——and Morley, M. H. (1981). 'The US Imperial State', in J. F. Petras (ed.), *Class, State, and Power in the Third World: with Case Studies on Class Conflict in Latin America.* Montclair and London: Allanheld, Osmun, and Zed Press: 1–36.

Phiraphol Tritasavit (1978). 'Labor Policy and Practices in Thailand: A Study of Government Policy on Labor Relations, 1932–1976', Ph.D. thesis, New York University.

Plai-Auw Chananont (1986). 'The Role of Merchant Capitalism in the Rise and Expansion of Capitalism in Northern Thailand, 1921–1980', Ph.D. thesis, Chulalongkorn University (in Thai).

Porpora, D. and Lim, M. H. (1987). 'The Political Economic Factors of Migration to Bangkok'. *Journal of Contemporary Asia*, 17/1: 76–89.

—— ——and Prommas, U. (1989). 'The Role of Women in the International Division of Labor: The Case of Thailand'. *Development and Change*, 20/2: 269–94.

Porter, P. W. and Sheppard, E. S. (1998). *A World of Difference: Society, Nature, Development.* New York: Guilford Press.

Potter, D. M. (1996). *Japan's Foreign Aid to Thailand and the Philippines.* New York: St. Martin's Press.

Poulantzas, N. A. (1975). *Classes in Contemporary Capitalism.* London: New Left Books.

Pradit Charsombut (1990). *Provincial Industry Labor Market.* Bangkok: Thailand Development Research Institute.

Pridi Banomyang (2000). *Pridi by Pridi: Selected Writings on Life, Politics, and Economy.* Chiang Mai: Silkworm Books.

Prizzia, R. (1985). *Thailand in Transition: The Role of Oppositional Forces.* Honolulu: University of Hawai'i Press.

Provincial Industry Office of Thailand (PIO) (1992–4). *Statistics of Provincial Industry* (by region). Bangkok: Ministry of Industry. (in Thai.)

Puey Ungphakorn (1977). 'Violence and the Military Coup in Thailand'. *Bulletin of Concerned Asian Scholars*, 9/3: 4–12.

Race, J. (1974). 'The War in Northern Thailand'. *Modern Asian Studies*, 8/1: 85–112.

Ramstetter, E. D. (1994). 'Comparisons of Japanese Multinationals and Other Firms in Thailand's Non-Oil Manufacturing Industries'. *ASEAN Economic Bulletin*, 11/1: 36–58.

Reid, A. (1988). *Southeast Asia in the Age of Commerce, 1450–1680*, i. *The Lands Below the Winds*. New Haven: Yale University Press.

——(1993). *Southeast Asia in the Age of Commerce, 1450–1680*, ii. *Expansion and Crisis*. New Haven and London: Yale University Press.

Rigg, J. (2003). *Southeast Asia: The Human Landscape of Modernization and Development* (2nd edn). London and New York: Routledge.

Riggs, F. W. (1966). *Thailand: The Modernization of a Bureaucratic Polity*. Honolulu: East–West Center Press.

Rock, M. T. (1995). 'Thai Industrial Policy: How Irrelevant Was it to Export Success?'. *Journal of International Development*, 7/5: 745–57.

Rodan, G. (1998). *The Political Economy of Singapore's Industrialization: National State and International Capital.* Basingstoke: Macmillan.

Ross, H. and Suwattana Thadaniti (1995). 'The Environmental Costs of Industrialization', in M. Krongkaew (ed.), *Thailand's Industrialization and its Consequences*. London, and New York: Macmillan and St. Martin's Press: 267–88.

Rostow, W. W. (1960). *The Stages of Economic Growth: A Non-Communist Manifesto*. Cambridge: Cambridge University Press.

Rotter, A. J. (1987). *The Path to Vietnam: Origins of the American Commitment to Southeast Asia*. Ithaca, NY: Cornell University Press.

Ruangrai Tokrisna and Panayotou, T. (1985). 'The Economics of Rural Livestock in Thailand', in T. Panayotou (ed.), *Food Policy Analysis in Thailand.* Bangkok: Agricultural Development Council, 197–221.

Sadao, F. (1989). 'Foreign Trade, Investment, and Industrial Imperialism in Postwar Japan', in T. Morris-Suzuki and T. Seiyama (eds.), *Japanese Capitalism Since 1945*. Armonk, NY and London: M. E. Sharpe: 166–206.

Saiyud Kerdphol (1985). *The Struggle for Thailand: Counter-Insurgency, 1965–1985*. Bangkok: S. Research Center Co.

Sanders, J. W. (1983). *Peddlers of Crisis: The Committee on the Present Danger and the Politics of Containment*. Boston: South End Press.

Saroj Aungsumalin (1990). *Finance, Credit and Provincial Industrialization*. Bangkok: Thailand Development Research Institute.

Sauwalak Kittiprapas (1999). 'Social Impacts of the Thai Economic Crisis', in *Social Impacts of the Asian Economic Crisis in Thailand, Indonesia, Malaysia and the Philippines*. Bangkok: Thailand Development Research Institute.

Schaller, M. (1985). *The American Occupation of Japan: The Origins of the Cold War in Asia*. New York: Oxford University Press.

Schmidt, J. D. (1996). 'Paternalism and Planning in Thailand: Facilitating Growth without Social Benefits', in M. Parnwell (ed.), *Uneven Development in Thailand.* Aldershot and Brookfield Vt.: Avebury: 63–81.

Scott, J. C. (1985). *Weapons of the Weak: Everyday Forms of Peasant Resistance*. New Haven and London: Yale University Press.

Seksan Prasertkul (1989). 'The Transformation of the Thai State and Economic Change (1855–1945)', Ph.D. thesis, Cornell University.

Sen, A. K. (1987). *The Standard of Living*. Cambridge and New York: Cambridge University Press.

——(1999). *Development as Freedom*. New York: Knopf.

Shaikh, A. and Tonak, A. (1994). *Measuring the Wealth of Nations: The Political Economy of National Accounts*. Cambridge and New York: Cambridge University Press.

Sheppard, E. S. and Barnes, T. J. (1990). *The Capitalist Space Economy: Geographical Analysis after Ricardo, Marx and Sraffa*. Cambridge, Mass.: Unwin Hyman.

Shoup, L. H. and Minter, W. (1977). *Imperial Brain Trust: The Council on Foreign Relations and United States Foreign Policy*. New York: Monthly Review Press.

Shuto, M. (1990). 'ODA in Japan's Foreign Policy: Changes, Characteristics and Prospects'. *Asian Review*, 4: 20–44.

Siegel, L. (1975). 'Asian Labor: The American Connection'. *Pacific Research & World Empire Telegram*, 6/5: 1–13.

Siffin, W. J. (1966). *The Thai Bureaucracy: Institutional Change and Development*. Honolulu: East–West Center Press.

Silcock, T. H. (1967*a*). 'Promotion of Industry and the Planning Process', in T. H. Silcock (ed.), *Thailand: Social and Economic Studies in Development*. Canberra: Australian National University Press in association with Duke University Press, 258–88.

——(1967*b*). 'The Rice Premium and Agricultural Diversification', in T. H. Silcock (ed.), *Thailand: Social and Economic Studies in Development*. Canberra: Australian National University Press in association with Duke University Press, 231–57.

Sims, B. (1992). *Workers of the World Undermined: American Labor's Role in U.S. Foreign Policy*. Boston: South End Press.

Skinner, G. W. (1957). *Chinese Society in Thailand: An Analytical History*. Ithaca, NY: Cornell University Press.

Sklair, L. (2001). *The Transnational Capitalist Class*. Oxford and Malden, Mass.: Blackwell.

Slater, D. (1989). *Territory and State Power in Latin America: The Peruvian Case*. New York: St. Martin's Press.

Smith, D. A. (1996). *Third World Cities in Global Perspective: The Political Economy of Uneven Urbanization*. Boulder, Colo.: Westview Press.

Smith, N. (1986). 'On the Necessity of Uneven Development'. *International Journal of Urban and Regional Research*, 10/1: 87–104.

——(1991). *Uneven Development: Nature, Capital and the Production of Space*. Oxford: Blackwell.

Somrudee Nicrowattanayingyong (1991). 'Development Planning, Politics, and Paradox: A Study of Khon Kaen, A Regional City in Northeast Thailand', Ph.D. thesis, Syracuse University.

Spooner, D. (1989). *Partners or Predators: International Trade Unionism and Asia*. Hong Kong: Asia Monitor Resource Center.

Standing, G. (1989). 'Global Feminization through Flexible Labor'. *World Development*, 17/7: 1077–95.

Steven, R. (1990). *Japan's New Imperialism*. Basingstoke: Macmillan.

——(1994). 'New World Order: A New Imperialism'. *Journal of Contemporary Asia*, 24/3: 271–96.

——(1996). *Japan and the New World Order: Global Investments, Trade, and Finance*. New York and Basingstoke: St. Martin's Press and Macmillan.

Stifel, L. D. (1976). 'Technocrats and Modernization in Thailand'. *Asian Survey*, 16/12: 1184–96.

Storper, M. and Walker, R. (1989). *The Capitalist Imperative: Territory, Technology, and Industrial Growth*. Oxford and New York: Blackwell.

Suehiro, A. (1989). *Capital Accumulation in Thailand, 1855–1985*. Tokyo: Centre for East Asian Cultural Studies.

Suganya Hutaserani and Somchai Jitsuchon (1988). 'Thailand's Income Distribution and Poverty Profile and their Current Situations' 1988 TDRI Year-End Conference on Income Distribution and Long-term Development. Paper. Bangkok: Thailand Development Research Institute.

Sukanya Hantrakul (1988). 'Prostitution in Thailand', in G. Chandler, N. Sullivan, and J. Branson (eds.), *Development and Displacement: Women in Southeast Asia*. Melbourne: Centre of Southeast Asian Studies, Monash University, 115–36.

Sungsidh Piriyarangsan (1983). *Thai Bureaucratic Capitalism, 1932–1960*. Bangkok: Social Research Institute, Chulalongkorn University.

——(1988). 'The Rise of Labor Movement in Thailand: An Analysis of Public Enterprise Workers'. *Asian Review*, 2: 54–78.

——(1989). 'The Foundation of a Workers Strategic Group', Ph.D. thesis, Bielefeld University.

——(1991). 'State-Enterprise Union: A Step Backward'. Working Paper no. 1. Bangkok: The Political Economy Centre, Faculty of Economics, Chulalongkorn University.

——and Kanchada Poonpanich (1994). 'Labour Institutions in an Export-Oriented Country', in G. Rodgers (ed.), *Workers, Institutions and Economic Growth in Asia*. Geneva: International Labor Organization: 211–53.

—— ——(1996). 'The Industrial Relations in Thailand', in Sungsidh Piriyarangsan and S. Itoga (eds.), *Industrial Relations System in Thailand*. Tokyo: Institute of Developing Economies.

Sunkel, O. (1973). 'Transnational Capitalism and National Disintegration in Latin America'. *Social and Economic Studies*, 22/1: 132–76.

Surachart Bamrungsuk (1988). *United States Foreign Policy and Thai Military Rule, 1947–1977*. Bangkok: Editions Duang Kamol.

Surin Maisrikrod (1997). 'The Making of Thai Democracy: A Study of Political Alliances Among the State, the Capitalists, and the Middle Class', in Anek Laothamatas (ed.), *Democratization in East and Southeast Asia*. Singapore, Chiang Mai, and New York: Institute of Southeast Asian Studies, Silkworm Books, and St. Martin's Press, 141–66.

Susumu, T. (2001). 'The Global Meaning of Japan: The State's Persistently Precarious Position in the World Order', in G. D. Hook and H. Harukiyo (eds.) *The Political Economy of Japanese Globalization*. London and New York: Routledge: 19–39.

Suthy Prasartset (1991). *Democratic Alternatives to Maldevelopment: The Case of Thailand*. Totsuka, Yokohama: PRIME (International Peace Research Institute Meigaku).

Tabb, W. K. (1995). *The Postwar Japanese System*. New York and Oxford: Oxford University Press.

Tanham, G. K. (1974). *Trial in Thailand*. New York: Crane Russak.

Taylor, P. J. (1994). 'The State as Container: Territoriality in the Modern World-System'. *Progress in Human Geography*, 18/2: 151–62.

——(1995). 'Beyond Containers: Internationality, Interstateness, Interterritoriality'. *Progress in Human Geography*, 19/1: 1–15.

Thak Chaloemtiarana (1979). *Thailand: The Politics of Despotic Paternalism*. Bangkok: Social Science Association of Thailand and Thai Khadi Institute, Thammasat University.

Theobald, S. (1995). 'Pressure Points in Industrial Development: A Gender Analysis Framework of the Various Stakeholders in the Northern Regional Industrial Estate, Thailand', MA thesis, University of East Anglia.

Thompson, R. (1966). *Defeating Communist Insurgency: The Lessons of Malaya and Vietnam*. New York: F. A. Praeger.

——(1968). 'Squaring the Error'. *Foreign Affairs*, 46/3: 442–53.

Thongchai Winichakul (1994). *Siam Mapped: A History of the Geo-Body of a Nation*. Honolulu: University of Hawai'i Press.

Thongnoi Keuakul (1983). 'A Study for Planning of Establishment of Northern Industrial Estate (Chiang Mai, Lamphun, Lampang)' MA thesis, Chulalongkorn University (in Thai).

Thorbek, S. (1987). *Voices from the City: Women of Bangkok*. London and Atlantic Highlands, NJ: Zed Books.

Tiano, S. (1994). *Patriarchy on the Line: Labor, Gender, and Ideology in the Mexican Maquila Industry*. Philadelphia: Temple University Press.

Tienchai Wongchaisuwan (1993). 'The Political Economy of Thailand: The Thai Peripheral State, 1958–1988', Ph.D. thesis SUNY-Binghamton.

Trocki, C. (1977). 'Boonsanong Punyodyana: Thai Socialist and Scholar, 1936–1976'. *Bulletin of Concerned Asian Scholars*, 9/3: 48–51.

Tronti, M. (1971). 'Social Capital', in *Operai e Capitale*. Turin: Einaudi, 267–311.

Truong, T.-a. (1990). *Sex, Money, and Morality: Prostitution and Tourism in Southeast Asia*. London and Atlantic Highlands, NJ: Zed Books.

Turton, A. (1978). 'The Current Situation in the Thai Countryside'. *Journal of Contemporary Asia*, 8/1: 104–42.

——(1984). 'Limits of Ideological Control and the Formation of Social Consciousness in South East Asia', in A. Turton and S. Tanabe (eds.), *History and Peasant Consciousness in South East Asia*. Osaka: National Museum of Ethnology.

——(1987). 'Rural Social Movements: The Peasants' Federation of Thailand', in A. Turton (ed.), *Production, Power, and Participation in Rural Thailand: Experiences of Poor Farmers' Groups*. Geneva: UN Research Institute for Social Development, 35–43.

——(1989*a*). 'Local Powers and Rural Differentiation', in G. P. Hart, A. Turton, and B. White (eds.), *Agrarian Transformations: Local Processes and the State in Southeast Asia*. Berkeley, Calif.: University of California Press, 70–97.

——(1989*b*). 'Thailand: Agrarian Bases of State Power', in G. P. Hart, A. Turton, and B. White (eds.), *Agrarian Transformations: Local Processes and the State in Southeast Asia*. Berkeley, Calif.: University of California Press, 53–69.

Unger, D. (1998). *Building Social Capital in Thailand: Fibres, Finance, and Infrastructure*. Cambridge and New York: Cambridge University Press.

United Nations Conference on Trade and Development (UNCTAD) (1997–2002). *World Investment Report*. Geneva: UNCTAD.
United Nations Development Program (UNDP) (1993–2000). *Human Development Report*. New York: Oxford University Press.
United Nations Industrial Development Organization (UNIDO) (1992). *Thailand: Coping with the Strains of Success*. Oxford and Cambridge, Mass.: Blackwell and UNIDO.
United Nations (UN), Statistical Office (1965–97). *International Trade Statistics Yearbook*. New York: UN.
US Agency for International Development (USAID) (1972). *Economic Assistance to Thailand*. Washington: USAID.
——(no date). Project database. Rosslyn, Va., USAID Library.
US Bureau of the Census (2000). *US International Trade Statistics Yearbook*. On-line edn. Available at http://www.census.gov/foreign-trade/www/press.html.
US Department of Labor (1959). *Foreign Labor Information: Labor in Thailand*. Washington: US Department of Labor.
——(1991–4). *Foreign Labor Information: Labor in Thailand*. Washington: US Department of Labor.
US Dept. of State. *Foreign Relations of the United States* (*FRUS*) (1950–60). Washington: US Government Printing Office.
Varunee Purisinsit and Benja Jirapatpimol (1994). 'Women's Industrial Work Conditions and the Changing of Family Relations'. Report submitted to the Hitachi Scholarship Foundation.
Vella, W. F. (1955). *The Impact of the West on Government in Thailand*. Berkeley, Calif.: University of California Press.
Vichote Vanno (1991). 'The Role of Trade Unions in the Political Development of Thailand, 1958–1986', Ph.D. thesis, City University of New York.
Voravidh Charoenloet (1989). 'The Crisis of State Enterprises in Thailand'. *Journal of Contemporary Asia*, 19/2: 206–17.
——(1996). 'A Study of Contract Labour in the Textile and Garment Sector in Thailand'. Unpub. MS. Bangkok: Faculty of Economics, Chulalongkorn University.
——(1998). 'The Situation of Health and Safety in Thailand'. Unpub. MS. Bangkok: Faculty of Economics, Chulalongkorn University.
Wade, R. (1990). *Governing the Market: Economic Theory and the Role of Government in East Asian Industrialization*. Princeton: Princeton University Press.
——(1996*a*). 'Globalization and Its Limits: Reports of the Death of the National Economy are Greatly Exaggerated', in S. Berger and R. Dore (eds.), *National Diversity and Global Capitalism*. Ithaca, NY and London: Cornell University Press, 60–88.
——(1996*b*). 'Japan, the World Bank, and the Art of Paradigm Maintenance: The East Asian Miracle in Political Perspective'. *New Left Review*, 217: 3–36.
Walby, S. (1990). *Theorizing Patriarchy*. Oxford: Blackwell.
Warr, P. G. (1993). 'The Thai Economy', in P. G. Warr (ed.), *The Thai Economy in Transition*. Cambridge and New York: Cambridge University Press, 1–80.
——(1999). 'What Happened to Thailand?' *World Economy*, 22/5: 631–50.
——and N. Bhanupong (1996). *Thailand's Macroeconomic Miracle: Stable Adjustment and Sustained Growth*. Washington: World Bank.

Wathinee Boonchalaksi and Guest, P. (1994). *Prostitution in Thailand.* Salaya, Phutthamonthon Nakhon Pathom, Thailand: Institute for Population and Social Research, Mahidol University.

Webber, M. J. and Rigby, D. L. (1996). *The Golden Age Illusion: Rethinking Postwar Capitalism.* New York: Guilford Press.

Wehmhörner, A. (1983). 'Trade Unionism in Thailand—A New Dimension in a Modernising Society'. *Journal of Contemporary Asia*, 13/4: 481–97.

Weiss, L. (1997). 'Globalization and the Myth of the Powerless State'. *New Left Review*, 225: 3–27.

Weisskopf, T. (1979). 'Marxian Crisis Theory and the Rate of Profit in the Postwar U.S. Economy'. *Cambridge Journal of Economics*, 3/4: 341–78.

Wilawan Tanratanakoon (1984). 'Linkage Patterns of Small Scale Industries in Chiang Mai Province', Rangsit: Asian Institute of Technology.

Williamson, H. (1994). *Coping with the Miracle: Japan's Unions Explore New International Relations.* London, Boulder, Colo., and Hong Kong: Pluto Press and Asia Monitor Resource Center.

Wittayakorn Chiengkul (1983). 'The Transformation of the Agrarian Structure of Central Thailand, 1960–1980'. *Journal of Contemporary Asia*, 13/3: 340–60.

Woo, J.-e. (1991). *Race to the Swift: State and Finance in Korean Industrialization.* New York: Columbia University Press.

Woo-Cumings, M. (1995). 'Developmental Bureaucracy in Comparative Perspective: The Evolution of the Korean Civil Service', in H.-K. Kim (ed.), *The Japanese Civil Service and Economic Development.* Oxford: Clarendon Press, 431–58.

——(1999). 'Introduction: Chalmers Johnson and the Politics of Nationalism and Development', in M. Woo-Cumings (ed.), *The Developmental State.* Ithaca, NY: Cornell University Press, 1–31.

World Bank (1959). *A Public Development Program for Thailand: Report of a Mission Organized by the International Bank for Reconstruction and Development at the Request of the Government of Thailand.* Washington: World Bank.

——(1975–2000). *World Bank Annual Report.* Washington: World Bank.

——(1980). *An Industrial Development Strategy for Thailand.* Washington: World Bank.

——(1993). *The East Asian Miracle: Economic Growth and Public Policy.* New York: Oxford University Press.

Wyatt, D. K. (1984). *Thailand: A Short History.* New Haven: Yale University Press.

Yeung, H. W.-c. (1998). 'Capital, State and Space: Contesting the Borderless World'. *Transactions of the Institute of British Geographers*, NS 23/3: 291–309.

Young, M. (1991). *The Vietnam Wars, 1945–1990.* New York: HarperCollins.

Yuangrat Wedel (1982). 'Current Thai Radical Ideology: The Returnees from the Jungle'. *Contemporary Southeast Asia*, 4/1: 1–18.

Zimmerman, C. C. (1931). *Siam Rural Economic Survey, 1930–31.* Bangkok: Bangkok Times Press.

INDEX